Sustainable Development Goals Series

World leaders adopted Sustainable Development Goals (SDGs) as part of the 2030 Agenda for Sustainable Development. Providing in-depth knowledge, this series fosters comprehensive research on these global targets to end poverty, fight inequality and injustice, and tackle climate change.

The sustainability of our planet is currently a major concern for the global community and has been a central theme for a number of major global initiatives in recent years. Perceiving a dire need for concrete benchmarks toward sustainable development, the United Nations and world leaders formulated the targets that make up the seventeen goals. The SDGs call for action by all countries to promote prosperity while protecting Earth and its life support systems. This series on the Sustainable Development Goals aims to provide a comprehensive platform for scientific, teaching and research communities working on various global issues in the field of geography, earth sciences, environmental science, social sciences, engineering, policy, planning, and human geosciences in order to contribute knowledge towards achieving the current 17 Sustainable Development Goals.

This Series is organized into eighteen subseries: one based around each of the seventeen Sustainable Development Goals, and an eighteenth subseries, "Connecting the Goals," which serves as a home for volumes addressing multiple goals or studying the SDGs as a whole. Each subseries is guided by an expert Subseries Advisor.

Contributions are welcome from scientists, policy makers and researchers working in fields related to any of the SDGs. If you are interested in contributing to the series, please contact the Publisher: Zachary Romano [Zachary.Romano@springer.com].

More information about this series at http://www.springer.com/series/15486

Evan Thomas

The Global Engineers

Building a Safe and Equitable World Together

Springer

Evan Thomas
Mortenson Center in Global Engineering
University of Colorado Boulder
Boulder, CO, USA

ISSN 2523-3084 ISSN 2523-3092 (electronic)
Sustainable Development Goals Series
ISBN 978-3-030-50262-1 ISBN 978-3-030-50263-8 (eBook)
https://doi.org/10.1007/978-3-030-50263-8

This Springer imprint is published by the registered company Springer Nature Switzerland AG
The registered company address is: Gewerbestrasse 11, 6330 Cham, Switzerland

Preface

The Global Engineers: Building a Safe and Equitable World Together, is inspired by both innovative and familiar opportunities for engineers to contribute to global prosperity. This book presents a vision for Global Engineering and identifies that engineers should be concerned with the unequal and unjust distribution of access to basic services, such as water, sanitation, energy, food, transportation, and shelter. As engineers, we should place an emphasis on identifying the drivers, determinants, and solutions to increasing equitable access to reliable services. Global Engineering envisions a world where everyone has safe water, sanitation, energy, food, shelter, and infrastructure, and can live in health, dignity, and prosperity.

This book seeks to examine the role and ultimately the impact of engineers in global development.

Engineers are solutions-oriented people. We enjoy the opportunity to identify a product or need, and design appropriate technical solutions. However, the structural and historical barriers to global prosperity require that engineers focus more broadly on improving the tools and practice of poverty reduction and that we include health, economics, policy, and governance as relevant expertise with which we are conversant. Engineers must become activists and advocates, leveraging our professional skills and capacity to generate evidence and positive impact toward rectifying inequalities and improving lives. Engineers must reject the ahistorical, technocratic and neo-colonial conceit that poverty can be solved through products or projects, or on a community scale that requires the poorest people to overcome historical and structural inequalities and injustices.

Half of this book is dedicated to profiles of engineers and other technical professionals who have dedicated their careers to searching for solutions to global development challenges. These professionals include Heather Fleming, a Navajo designer and the founder of Catapult Design, Chantal Iribagiza, a Rwandese engineer specializing in rural water supplies, Jean Ntzinda, a Rwandese environmental planner responsible for facilitating foreign corporate and donor-supported programs, Avery Bang, an American civil engineer and CEO of Bridges to Prosperity, Doris Kaberia, a Kenyan expert in food security and pastoral livelihoods, Petros Birhane, an Ethiopian agricultural engineer and disaster relief expert, and Dan Hollander, an American hydrological engineer and former foreign service officer. These stories introduce the reader to the diverse opportunities and challenges in Global Engineering.

The Global Engineers was finalized and published during the global COVID-19 coronavirus pandemic. Far from diminishing the relevance of Global Engineering, this global challenge amplifies the urgency of global cooperation to address the causes and create solutions to chronic and emerging environmental challenges—climate change, diseases, droughts, floods, clean air, water, sanitation, hygiene, and safe and reliable infrastructure.

COVID-19 has caused social and economic disruption worldwide and a reckoning of the balance in our priorities between public health and the economy. Our global response is heartening, and demonstrates our potential to work collectively to protect each other. Yet, the public health impact and economic effects of this pandemic will not be borne equally despite these collective measures. Devastatingly, UNICEF estimates that an additional 6,000 children could die from preventable causes in 2020 alone as COVID-19 disrupts health systems and routine services. "Under a worst-case scenario, the global number of children dying before their fifth birthdays could increase for the first time in decades," said UNICEF Executive Director Henrietta Fore. "We must not let mothers and children become collateral damage in the fight against the virus. And we must not let decades of progress on reducing preventable child and maternal deaths be lost." Further, the United Nations International Labour Organization estimates that nearly half of the global workforce have lost their jobs, including 1.6 billion of the 2 people who work in informal jobs. The United Nations University estimates that as many as half a billion people could be pushed back into poverty because of the economic impacts of COVID-19, the first such increase in global poverty since 1990. The World Bank forecasts that sub-Saharan Africa will face a severe food security crisis and dramatic increases in unemployment. In some ways, we are not in this together: COVID-19 will be exacerbated by underlying, chronic inequalities as basic as access to safe water, sanitation and hygiene (WASH). This intersectionality—where these historic inequalities in access to basic services may accelerate the spread of COVID-19—deserves attention equal to our present emergency response.

While we don't yet know how many people will die from COVID-19, we do know that an estimated 842,000 people die every single year from a lack of safe drinking water and inadequate sanitation and hygiene. These deaths are almost entirely preventable—especially if we collectively made the extraordinary investments in solving these chronic problems that we are taking now with COVID-19.

The World Health Organization (WHO) has provided clear guidance on the most critical measure we can all take to protect human health and reduce the spread of COVID-19: *"Hands should be washed with soap and running water."* While clear and simple, this directive is far from attainable for the three billion people around the world who lack access to soap and safe water at home. 40% of the world's population does not have access to soap and water to wash their hands. In rural East Africa, only about 50–60% of households have improved water sources, and only 4–10% have a place to wash their hands with soap and water.

Claire Chase and Luis Andres, economists in the water practice at the World Bank, put it simply: *"Resources are also needed during a pandemic response to ensure that WASH services continue to function. The drastic effects of supply chain disruptions, depressed economic activity, and even panic-buying can worsen the quality and continuity of water and sanitation services. Whether through financial support to utilities, making treatment chemicals or fuel available, providing free WASH services to households and institutions that are most in need, or ensuring that soap and disinfectant are available, there are many interventions that can limit the spread of disease."*

The global community is responding to the COVID-19 pandemic, but in the process we are in danger of pulling resources away from addressing other chronic risks. The United States has even asked global partners to ship masks, gloves, and ventilators previously deployed to support humanitarian and development response, back to the United States. And yet, other human health hazards—like climate change, drought, food, and water security—have not gone away. We must apply our ability to act collectively to solve these chronic public health challenges.

Together, engineers can work alongside countries, communities, and other professionals to identify and dismantle the underlying causes of persistent global poverty, and elevate all people, and their environment, as Global Engineers.

Boulder, USA — Evan Thomas

Acknowledgments

This book is dedicated to Lauren Alstot and Desmond Thomas. The author thanks his colleagues, mentors, and friends who have contributed to the vision and work described in this book, including Mort, Alice and Dana Mortenson, Bernard Amadei, Amy Javernick-Will, Karl Linden, Rita Klees, Carlo Salvinelli, Laura MacDonald, Balji Rajagopalan, Dave Klaus, Bobby Braun, Keith Molenaar, Doug Smith, Eleanor Allen, Alex Dehgan, DeAndra Beck, Andrew Reynolds, Laura Brunson, Keith Wright, John Butterworth, Iana Aranda, Dan Irwin, Ashutosh Limaye, Emily Adams, Robinson Mugo, Jenny Frankel-Reed, Kevin Mulligan, Luis Andres, Crystal Fenwick, Richard Sedlmayr, Donna Dalton, Sarah Goodroad, Jamie McDevitt-Galles, Amy Hill, Jason Neff, Styvers Kathuni, Christian Muragijimana, Lambert Mugabo, Taylor Sharpe, Emily Bedell, Katie Fankhauser, Abby Bradshaw, Matthew Falcone, Anna Libey, Pranav Chintalapati, Denis Macharia, Rob Hope, Patrick Thompson, Johanna Koehler, Nick Turman-Bryant, Danny Wilson, Skot Croshere, Jeremy Coyle, Bud Pope, Carmel and Ron Garan, Niko Kalinic, Josh Kefauver, Kyle Silon, Victor Bernstein, Marshall Davert, John and Jodi Graf, Christina Barstow, Corey Nagel, Thomas Clasen, Miles Kirby, Alex Johnson, Andrea Johnson, Steve Alstot, Chuck Waldron, David Thomas, Bryan Thomas, and Eva Leidman; the subjects of the profiles in this book including Heather Fleming, Chantal Iribagiza, Jean Ntzinda, Avery Bang, Doris Kaberia, Petros Birhane and Dan Hollander. This book owes a debt to Encountering Poverty by Ananya Roy, Genevieve Negron-Gonzales, Kweku Opoku-Agyemang, and Claire Talkwaker, and The Divide by Jason Hickel.

Contents

About the Author

Evan Thomas is the Director of the Mortenson Center in Global Engineering and holds the Mortenson Endowed Chair in Global Engineering at the University of Colorado Boulder. He is a tenured Associate Professor jointly appointed to the Civil, Environmental, and Architectural Engineering and the Aerospace Engineering Sciences Departments. Evan is currently a member of the NASA and USAID SERVIR Applied Sciences Team, applying NASA satellite data with USAID supported expertise toward drought resilience in East Africa. Evan is the co-editor and co-author of the 2020 UNESCO engineering report, "Engineering the Sustainable Development Goals."

Evan has worked professionally in over a dozen countries. The Mortenson Center supports hundreds of undergraduate and graduate students working in over 20 countries, with dozens of nonprofit, social enterprise, government, and community partners.

Evan has a Ph.D. in Aerospace Engineering Sciences from the University of Colorado Boulder, is a registered Professional Engineer in the State of Texas, and has a Master's in Public Health from the Oregon Health and Science University.

Evan's technical background is in water and air testing and treatment. He founded SweetSense Inc., which is supported by USAID and the National Science Foundation to develop and apply satellite-connected sensors to monitor drinking water systems. The team monitors millions of people's water supply across east Africa on a daily basis.

Evan's research has been funded by NASA, the National Science Foundation, the World Bank, USAID, the Moore Foundation, the UN Foundation, the CDC, UNESCO, the United Kingdom Department for International Development, the Gates Foundation, and others.

From 2012 to 2016, as Chief Operating Officer of DelAgua Health, Evan designed and operated a $25 million public health intervention with the Government of Rwanda. The program supplied 350,000 households with cookstoves and 102,000 households with water filters in over 7,500 villages, reaching 1.6 million people.

From 2010 to 2018, Evan was a member of faculty at the Portland State University and Oregon Health & Science University, as well as founder of SweetLab and a founding director of GlobalPDX. In 2017, Evan was a finalist for the Canadian Astronaut Selection.

Evan was a civil servant at the NASA-Johnson Space Center in Houston, Texas from 2004 to 2010. At NASA, Evan worked as an aerospace engineer designing microgravity fluid management and water recovery systems for spacecraft hardware on the Space Shuttle and the International Space Station.

What Is Global Engineering?

1

Abstract

This chapter presents the field of Global Engineering, and identifies that engineers should be concerned with the unequal and unjust distribution of access to basic services, such as water, sanitation, energy, food, transportation, and shelter, and as engineers we should place an emphasis on identifying the drivers, determinants, and solutions to increasing equitable access to reliable services. Engineers must become activists and advocates, leveraging our professional skills and capacity to generate evidence and positive impact toward rectifying inequalities and improving lives. Engineers must reject the ahistorical, technocratic and neo-colonial conceit that poverty can be solved through products or projects, or on a community scale that requires the poorest people to overcome historical and structural inequalities and injustices. Global Engineering envisions a world where everyone has safe water, sanitation, energy, food, shelter, and infrastructure, and can live in health, dignity, and prosperity. This chapter adapts and updates the 2019 publication in *Sustainability, "Toward a New Field of Global Engineering"* (Thomas 2019).

1.1 Introduction

Engineers are solutions-oriented people. We enjoy the opportunity to identify a product or service need and design appropriate technical solutions. This model can be effective in high-income regions where the engineering profession is complemented by communities with strong political capital and tax bases leveraged to provide essential government services such as water, sanitation, electricity, and roads; an enforced regulatory environment to maintain the quality and safety of these services; and business and consumer markets to purchase products and services. Such necessary social supports are often invisible to the engineer, whose education does not typically include crash courses in economics or policy. As a result, engineers are poorly equipped to address or even recognize the existence of structural gaps to providing public and private services in lower income settings.

1.2 The World Today

There are abundant reasons to be optimistic about global development, public health, and poverty reduction. The billions of people who

E. Thomas, *The Global Engineers*, Sustainable Development Goals Series,
https://doi.org/10.1007/978-3-030-50263-8_1

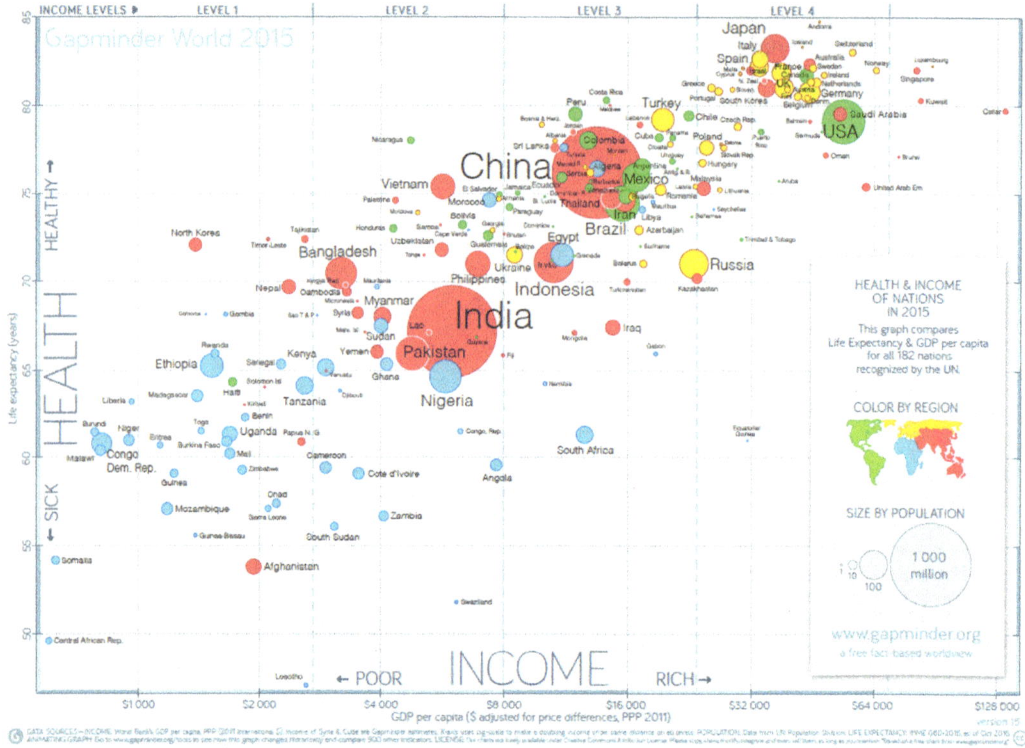

Fig. 1.1 GapMinder (www.gapminder.org) shows income levels (1–4) compared against life expectancy. Over the past 200 years, nearly every country and region in the world has progressed from the lower left toward the upper right, increasing in wealth and health. However, there remain clear disparities, highlighted by the concentration of countries in sub-Saharan Africa clustered in the lower left quadrant

have been vaccinated, the increasing number of people entering the middle class, and the myriad accomplishments cited in reports on the United Nations Millennium Development Goals (MDGs) and Sustainable Development Goals (SDGs) all testify to the positive impact of current development policies and practice on economic growth, life expectancy, and overall prosperity (Rosling et al. 2018).

Global progress has been unambiguously impressive. Over the past 200 years, nearly every country and region in the world has progressed from poverty and low life expectancy to better financial and health outcomes (Fig. 1.1). Likewise, global mortality among children five years old and younger (an important indicator of overall national public health) has shown dramatic declines for all major causes, including malaria, HIV/AIDS, respiratory infections, diarrheal diseases, neonatal complications, and birth defects (Fig. 1.2). These clear trends can be interpreted optimistically as inevitable global progress.

However, while there is clear, unambiguous progress toward greater wealth, health, and prosperity globally, most of this success over the past 30 years has occurred in south and east Asia, and has been dominated by economic growth in India and China, while sub-Saharan African countries continue to show low income levels (Fig. 1.1), underlining the need to question if current global development policies and efforts are effective at reducing poverty and improving health globally.

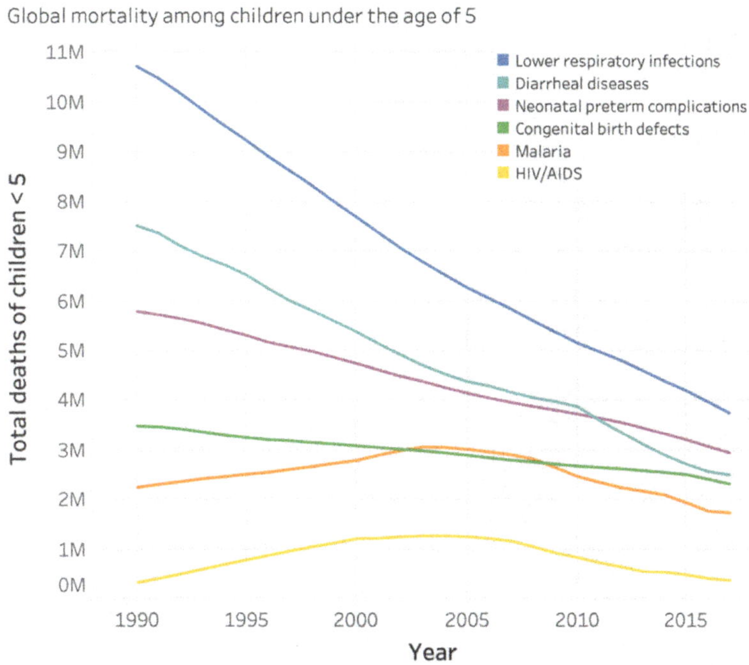

Fig. 1.2 Global reductions in mortality among children aged under 5 over the past 30 years (data from www.ourworldindata.org and www.healthdata.org)

Today, over half the world's population still lives on less than $5.50 a day (The World Bank 2018). The burden of disease in low-income countries is overwhelmingly attributable to environmental health issues including air quality and quality, sanitation, and disease vectors including malaria-carrying mosquitos (Institute for Health Metrics and Evaluation 2019). While the fraction of the world's population living in absolute poverty has decreased over the past 50 years, the absolute number of people in poverty—about 1 billion—has not changed over the last 30 years (Fig. 1.3).

Most internationally funded development efforts focus on sub-Saharan Africa—home to over a billion people—however, the number of people in this region in extreme poverty has only increased over the same period. Indeed, the World Bank projects that, by 2030, about 500 million people will live in extreme poverty, with the majority living in sub-Saharan Africa (Fig. 1.3). Of the 47 countries listed by the United Nations as "least developed," 33 are in sub-Saharan Africa—over 70% (UNFCCC 2020). The only least developed country (LDC) in the Americas is Haiti.

If a child under the age of 5 dies, there is a 99% chance that she was born outside of Europe or North America (UNICEF 2018). At the current rate of development, and assuming the same economic policies, it will take over 200 years for everyone in the world to earn at least $5 per day. Furthermore, achieving this level of growth will require a 175-fold increase in global production and consumption of resources compared to 2010 levels (Woodward 2015). The link between resource use and poverty is starkly illustrated in Fig. 1.4, which shows the per-person carbon dioxide emissions for every country in the world, and Fig. 1.5, which shows the global burden of disease per 100,000 people. These maps are practically mirror images of each other, with almost no energy use per person and dramatic disease burdens in sub-Saharan Africa.

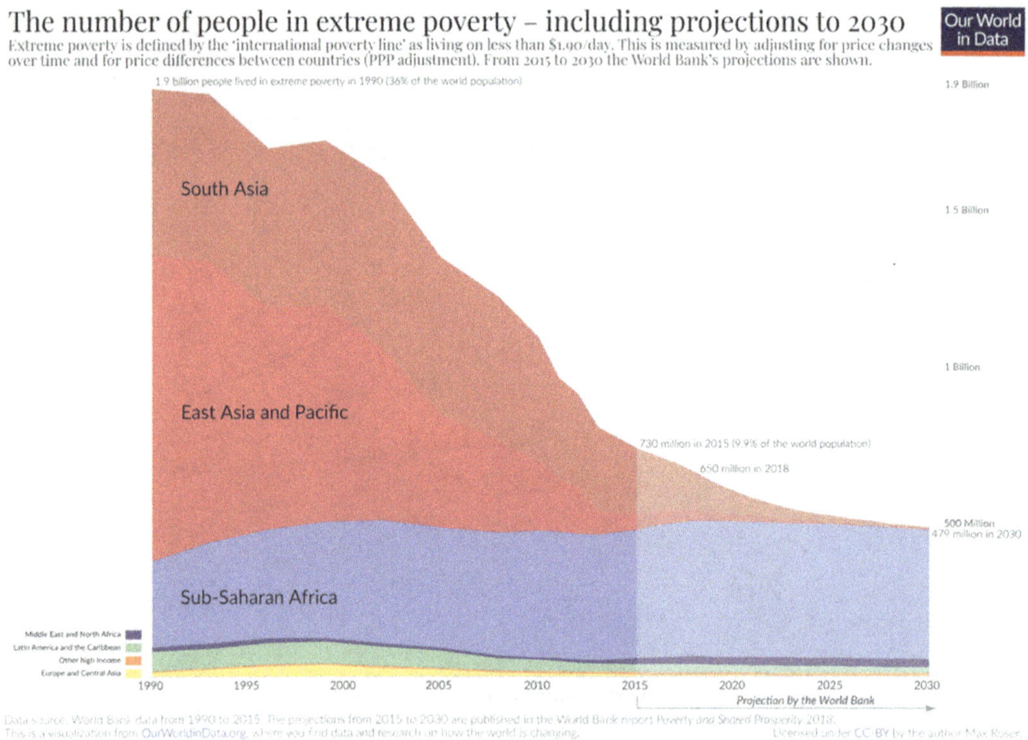

Fig. 1.3 Number of people in extreme poverty by region from 1990 with projections to 2030. Note the nearly stable number of people in extreme poverty in sub-Saharan Africa (www.ourworldindata.org)

Foreign aid and philanthropy are promoted as part of the solution to these chronic challenges. Indeed, over $160 billion per year is provided by high-income countries to low-income countries in part to address these conditions (World Bank 2019). Furthermore, tallying up all financial resources including aid, foreign investments, trade, debt cancelation, and remittances, over $2 trillion was transferred to developing countries in 2012, the last year for which a full dataset is available (Centre for Applied Research 2015).

So why are poverty and its associated conditions of contaminated water and air, rural isolation, and lack of energy access so persistent? One startling clue is that financial outflows—resources provided from low-income countries to high-income countries—dramatically exceed inflows. Financial outflows, attributable to debt interest payments (on World Bank and International Monetary Fund loans, much of which were undemocratically imposed on developing countries in the 1980s through Structural Adjustment Loans, now approaching $8 trillion dollars), repatriation of corporate profits, and capital flight including trade misinvoicing and tax avoidance, accounted for over $5 trillion in 2012 (Centre for Applied Research 2015). Furthermore, unfair trade agreements that restrict lower income countries from protecting their own industries perpetuates poverty. A grotesque example recently occurred in a trade dispute between the Republic of Rwanda and the United States. As reported in the Washington Post in March 2018, Rwanda, seeking to protect their own textile industry, imposed an import tax on used, hand-me-down clothing from the United States. In response, the Trump Administration slammed Rwanda with tariffs on their own apparel exports, an entirely legal move under the perversely named African Growth and Opportunity Act (AGOA), wherein "participating countries are required to adhere to a market economy

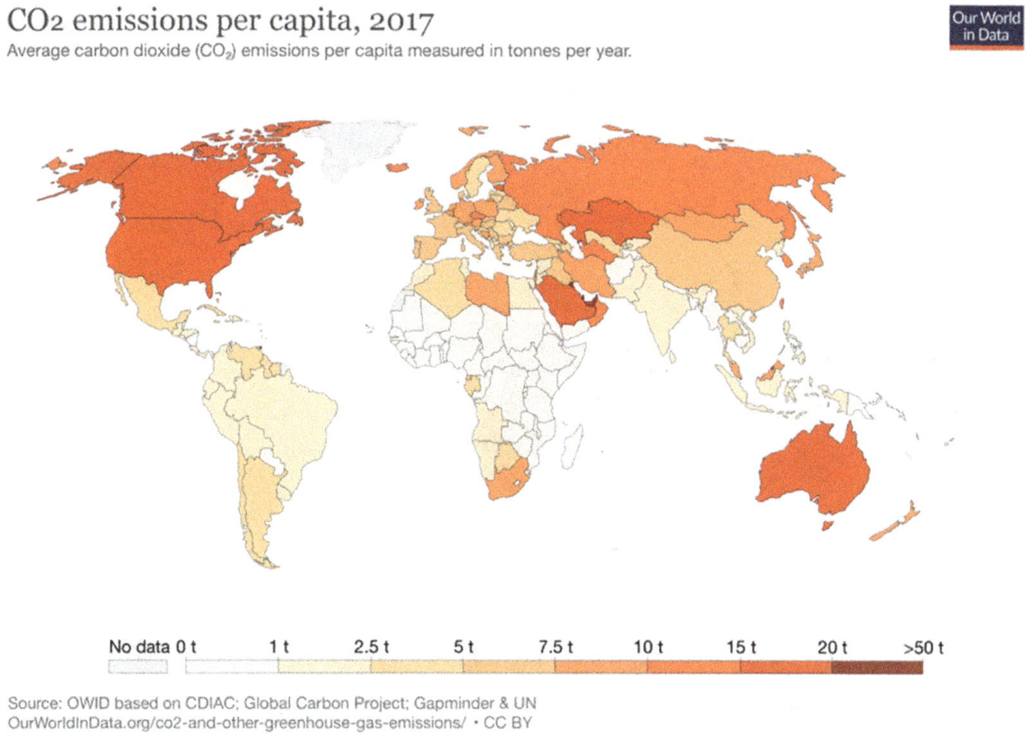

Fig. 1.4 Carbon dioxide emissions per capital for each country in the world in 2017. The United States remains the world's largest emitter of carbon per capita, while energy use per person in sub-Saharan Africa barely registers (www.ourworldindata.org)

and move toward the "elimination of barriers to U.S. trade and investment."

This $3 trillion per year in total net outflows represents 18 times the annual global foreign aid budget. In sub-Saharan Africa, global net financial flows account for approximately $20 billion per year (Fig. 1.6), roughly the annual budget of the National Aeronautics and Space Administration (NASA).

In other words—low-income countries are net creditors to rich countries. Income inequality on this global scale is also felt within rich countries and by individuals. Oxfam reported in 2017 that the richest eight people have more wealth than the poorest half of the world's population (OXFAM 2017). While many of these individuals are generous with their philanthropy, the global concentration of wealth is increasing.

Beyond even the chronic barriers to poverty reduction, climate change is exacerbating and accelerating poverty in some regions of the world. The World Health Organization (WHO) conservatively estimates that climate change-driven increases in temperature (heat waves), diarrhea, malaria, and malnutrition (crop failure) will result in over 250,000 additional deaths each year between 2030 and 2050 (World Health Organization 2018). A further 100 million people could be pushed back into poverty by 2030 because of climate change (Haines and Ebi 2019). Most of these deaths and hardships will occur in developing countries, which are among the populations least responsible for climate change and least equipped to manage its impacts.

In this global context, it is not surprising that engineers, while motivated to make a positive impact, are poorly trained and equipped to address these structural barriers to poverty reduction, which can overwhelm the contributions of engineered products or services.

Put simply—no new water filter product, social business sanitation service, elementary

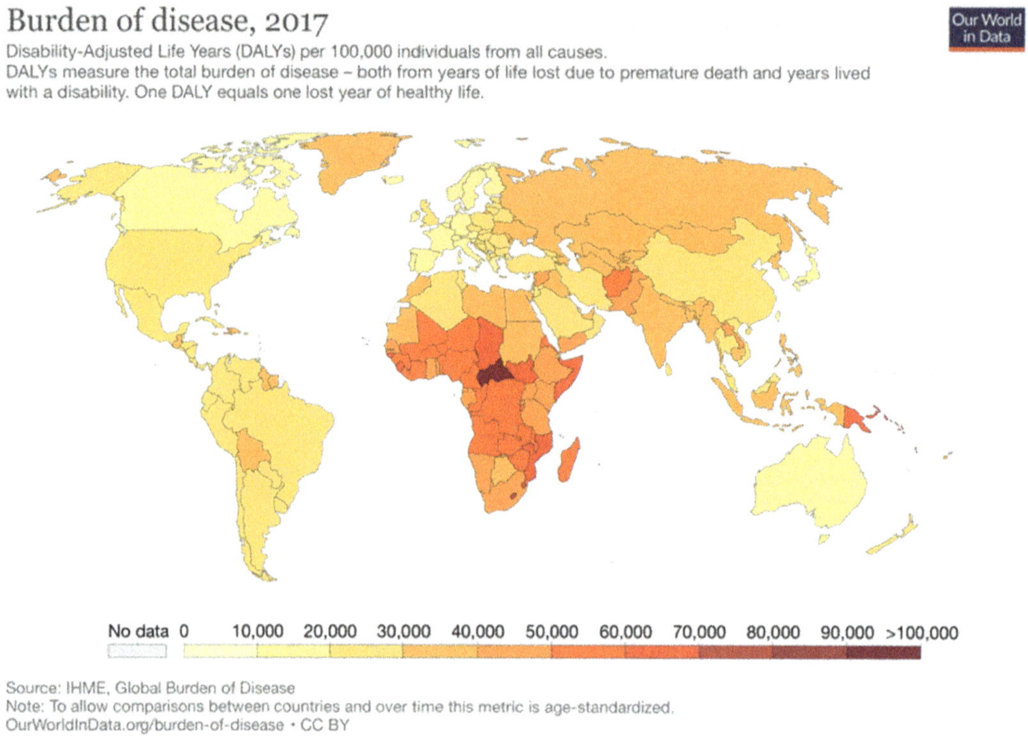

Fig. 1.5 The global burden of disease in Disability-Adjusted Life Years per 100,000 people, for 2017. The burden of disease is concentrated in sub-Saharan Africa. This chart appears to present the inverse picture of global per capita carbon dioxide emissions (www.ourworldindata.org)

school rainwater catchment tank, electricity grid, or training for a local government on the importance of water pump maintenance will make a dent in a system that precludes countries from developing empowered citizenry and robust tax bases that can support accountable and responsive governmental services.

1.3 The Engineer in the Past

The Engineer's first foray into the modern global development sector took the form of large-scale, top-down infrastructure such as electricity grids, dams, roadways, and water management systems, often implemented in former colonies after World War II. The perceived failure of this model, reflected in crumbling infrastructure in the 1970s, resulted in a pivot toward smaller scale engagement, based on community-level participatory development.

The response of engineering to community participatory development was provided by E. F. Schumacher, a British economist who coined the term "appropriate technology" in his book *Small is Beautiful*, published in 1973. An appropriate technology, according to Schumacher, is one that is small in scale, uses local materials, is energy efficient, environmentally sound, labor intensive, controlled by the community, and maintained locally. This well-intentioned approach sought to bypass the failures of large-scale development. Small-scale participatory development and appropriate technology models were adopted by many development agencies, and promoted as the most impactful and

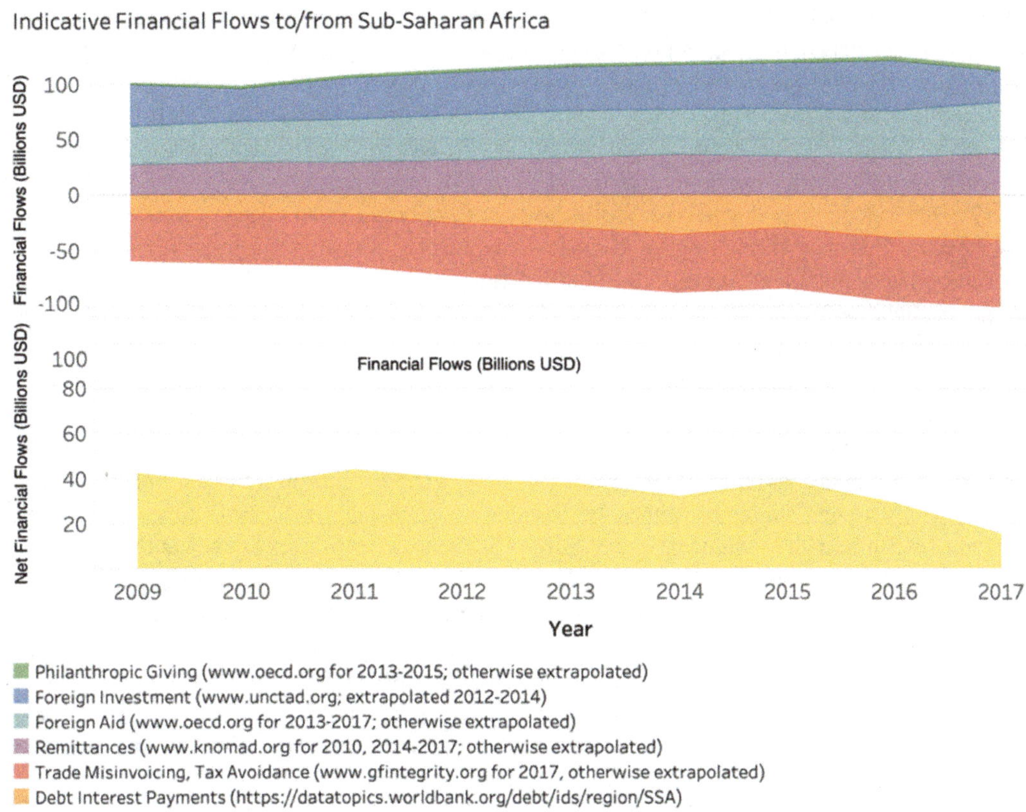

Fig. 1.6 Indicative financial flows to and from sub-Saharan Africa (2009–2017). Inflows include philanthropic giving, foreign direct investment, foreign aid, and remittances, with trade misinvoicing, tax avoidance, and debt interest payments as financial outflows

appropriate way to train engineers to engage in poverty reduction efforts (Amadei et al. 2009; Thomas and Amadei 2010).

These models, often referred to as participatory development, sustainable development, or appropriate technology, have an unfortunate common denominator—an assumption on the part of development funders, agencies, professionals, and volunteers that emerging economies are either not entitled to, or not capable of, benefiting from the same technologies as the countries providing development advice.

Such fashionable approaches expect developing communities to somehow grow their economies, reduce disease burdens, educate a population, and engage in global trade, all without the benefit of carbon, plastics, or professional specialization. Even if these approaches are somehow viable, the development organizations and individuals promoting them rarely come from communities or countries that reflect these modalities.

Which is more appropriate—a water filter made out of locally procured, low-cost clay that employs local residents in producing, selling, and maintaining the filters, or an ultrafiltration membrane and plastic contraption produced in China?

This sounds like an easy choice when viewed through the appropriate technology and community development lens. But what if the local ceramic filter has never been shown conclusively to improve health, while the imported Chinese filter has a strong track record in the epidemiology literature of improving the health of its recipients? Which choice would be more appropriate?

Or what about a locally produced stove built with local clay that has an emission profile no different from an open fire? Is it more appropriate than a clean, imported Chinese stove? Is a 500-watt solar panel really more appropriate than a similarly priced 2,000-watt diesel generator that is obtained through a preexisting supply chain?

A strict definition of appropriate technology can result in pitfalls when rigorous, multidisciplinary trade evaluations are neglected. While the local stove or filter may check 9 out of 10 boxes for "appropriate technology," it may fail to fulfill the fundamental purpose of the solution—to improve health. Recognizing this requires a simultaneous respect for public health, business, policy, and engineering expertise.

Today, a development agency would not think twice about promoting mobile technologies and off-grid solar energy, although neither fit the definition of appropriate technology. More current debates revolve around the appropriateness of importing higher quality, lower cost often-Chinese made products such as water filters and cookstoves, at the expense of local producers; or the effectiveness of giving away these kinds of health products versus charging consumers an (often subsidized) fee.

Unfortunately, over almost 50 years, appropriate technology and community participatory development has failed to eliminate or even substantially reduce the number of people living in poverty in developing countries. Meanwhile, the world has shrunk and technologies have advanced.

The terms "global engineer" (Amadei 2014), "development engineering" (Nilsson et al. 2014), "humanitarian engineering" (Mitcham and Munoz 2010), and "peace engineering" (Amadei 2019) have recently entered the lexicon of academia and professional disciplines.

Since 2001, Engineers Without Borders-USA (EWB-USA) has involved engineering students and professionals in extracurricular and volunteer engagement in developing communities. This approach has been recognized as an important component of professional training (Bourn and Neal 2008). However, there is increasing recognition that this approach is insufficient to train globally responsible engineers (Mintz et al. 2014), and that rigor equal to any other engineering discipline should be introduced at the curriculum level with engineers cross-trained in established development disciplines such as global health, economics, public policy, and social business (Nilsson et al. 2014).

Human-centered design and product and service development, often linked to social business models, have also been promoted as new approaches updating the role of the engineering profession in global development. In 2014, experts at the University of California Berkeley coined the term "Development engineering," promoting a model that links human-centered design, multidisciplinary teams, and user- and community-centric engagement toward product and service design (Nilsson et al. 2014). Acknowledging the limitations of mono-disciplinary approaches, the authors advance the premise that development engineering builds "on techniques from engineering, development economics, behavioral science, and sociology," and designs products and services on behalf of developing countries, while addressing market barriers and institutional failures, and promoting business models.

This product-centered focus can have complex outcomes. While evangelizing a community lens, well-intentioned entrepreneurs have released products that fail to address the symptoms or the causes of the issues they highlight, and instead attract resources to gimmicky solutions. Examples include the Play Pump, a merry-go-round water pump that was supposed to offer children a fun way to pump water and in reality often required women to walk the pump in circles (Stellar 2010), and the One Laptop per Child initiative, which suggested that a cheap, no-frills laptop could be produced cost effectively for low-income classrooms (Robertson 2018), but resulted in a dramatic decrease in mass-produced, higher quality and functioning consumer laptops and tablets.

Furthermore, many engineering efforts on a community or product scale have required either volunteer or low-salaried engineering labor, which has the effect of reducing the professional depth of the contributions of engineers to global development. Meanwhile, larger scale infrastructure contracts in the United States for work in low- and middle-income countries are awarded to major engineering and technical contractors (e.-g., AECOM, CH2MHill, TetraTech, Chemonics) that offer competitive salaries, but may not be mandated or capable of addressing longer term systemic development challenges.

An additional chronic limitation is the under-representation of engineers from low- and middle-income countries, with development programs often relying instead on short-term engagements by Western engineers.

While the literature contains laudatory case studies and attractive examples of successful products and services, in reality these are piece-meal patches to endemic structural challenges. Moreover, perceptions differ across professions: those involved in public health may think that the problem lies less with engineering than with behavior change; entrepreneurs often believe that failure to charge people for a product misses the mark; engineers are baffled by the "touchy-feely" aspects of development; and those working in public policy wonder how to manage unfunded mandates.

Heather Fleming, a Stanford-educated product designer, has struggled with these dynamics in her own career, working in Africa with funding from international donors, and now focusing on building entrepreneurial services at home on the Navajo reservation in the United States. Heather's story is shared in Chap. 4 of *The Global Engineers.*

1.4 The Engineer of the Future

The role of engineers in contributing to global poverty reduction and the SDGs requires an upgrade. The 2020 UNESCO report *Engineering the Sustainable Development Goals* frames this imperative: *"The engineering profession must embrace a new mission statement – to contribute to the building of a more sustainable, stable and equitable world... Together, we envision a world where all people have access to the services and resources necessary to live healthy, fulfilling lives and live in dignity and at peace, while working to preserve our global environment upon which we all depend. To do so, a new action-based blueprint for global engineering education, life-long education and practice is needed for the engineering profession to contribute to meeting the SDGs by 2030"* (Amadei and Thomas 2020).

Luckily, several complementary academic and professional fields, including Global Health and Development Economics, have a longstanding track record and philosophy from which we can learn.

Global Health as a field of study, research, and practice, is well established. The Consortium of Universities for Global Health describes it as a field that, "emphasizes transnational health issues, determinants, and solutions; involves many disciplines within and beyond the health sciences and promotes interdisciplinary collaboration; and is a synthesis of population-based prevention with individual-level clinical care." Moreover, Global Health "places a priority on improving health and achieving equity in health for all people worldwide."

Similarly, Development Economics, as embodied by the World Bank, is a field dedicated to studying and leveraging economic tools including taxes, trade, transfers, loans, and investment to improve economic growth in low-income countries.

The Global Health and Development Economics communities are grounded in method and tool development, evidence generation, and translation of findings into national and global policies. Many implemented policies, impact evaluations, or research studies conducted by economics or public health professionals are designed to address immediate needs and generate evidence to inform policies and funding decisions. Publications, methods, tools, and technologies are evaluated, tested, and refined, and consensus is built through meta-analysis and dissemination.

However, Global Health and Development Economics are imperfect models. Implementation and evaluation procedures are often designed and conducted with foreign funding and foreign experts, which can have the effect of reinforcing autocracies and creating a "tyranny of experts" (Peet 2014). Yet, the professionalization of these fields has resulted in a high degree of influence both on policy and the public.

This book presents the case that Global Engineering should be concerned with the unequal and unjust distribution of access to basic services, such as water, sanitation, energy, food, transportation, and shelter, and as engineers we should place an emphasis on identifying the drivers, determinants, and solutions to increasing equitable access to reliable services. Global Engineering envisions a world where everyone has safe water, sanitation, energy, food, shelter, and infrastructure, and can live in health, dignity, and prosperity.

Global Engineering can be the professional and academic complement to Global Health and Development Economics. It focuses on broadly improving the tools and practice of poverty reduction, and includes health, economics, policy, and governance as relevant dimensions, requiring professional engineers to be conversant in these fields. Global Engineers must also become activists and advocates, leveraging our professional skills and capacity to generate evidence and positive impact toward rectifying inequalities and improving lives. Engineers must reject the ahistorical, technocratic and neocolonial conceit that poverty can be solved through products or projects, or on a community scale that requires the poorest people to overcome historical and structural inequalities and injustices.

1.5 Fields of Global Engineering Practice

The role of the engineer in addressing today's global poverty challenges must be elevated. While village-scale interventions may have a positive impact on a community, product design may address some consumer demands, and large-scale infrastructure can in the short-term fill gaps in basic services, the structural constraints that perpetuate poverty require structural solutions.

It is challenging to strike a balance between forgiving optimism and paralyzing pessimism when examining the spectrum and arc of global development. As engineers, we want to be able to design and implement durable solutions. However, we need to broaden our perspective to include solutions to underlying structural problems.

The field of Global Engineering can contribute to addressing these structural issues, by developing and validating methods, tools, and standards that are broadly leveraged to increase poverty reduction. Technology development and demonstration, data collection, and impact evaluation can all contribute to evidence-based influence on policies and practice. Remote sensing technologies are informing conversations about the impacts of global warming; data collection and analysis technologies support impact evaluations by generating robust findings on the effectiveness of interventions; systems engineering is expanding the engineer's lens to more broadly consider institutions, governance, and financial planning in basic service delivery; and engineering education is embracing history, public health, and policy.

Global engineers must be taught to consider the historical and present causes of persistent poverty, instead of perceiving poverty as a stage of inevitable growth that can be helped along with conventional technical solutions. Such training will better inform the choices engineers make and help move the engineering sector away from a product and village-level focus toward working to address the root causes of poverty. Unequal distribution of wealth and resources, and continued exploitation of low-income communities by the global economic system, will undermine the effectiveness of small-scale programs and products.

The field of Global Engineering, while working toward improving policies and equity globally, need not be divorced from innovating solutions and designing technologies. Indeed, technological innovations can support knowledge generation, policy, and public participation in acknowledging and addressing the root causes of persistent poverty. This section provides several pertinent, but by no means exhaustive, examples from a variety of institutions and

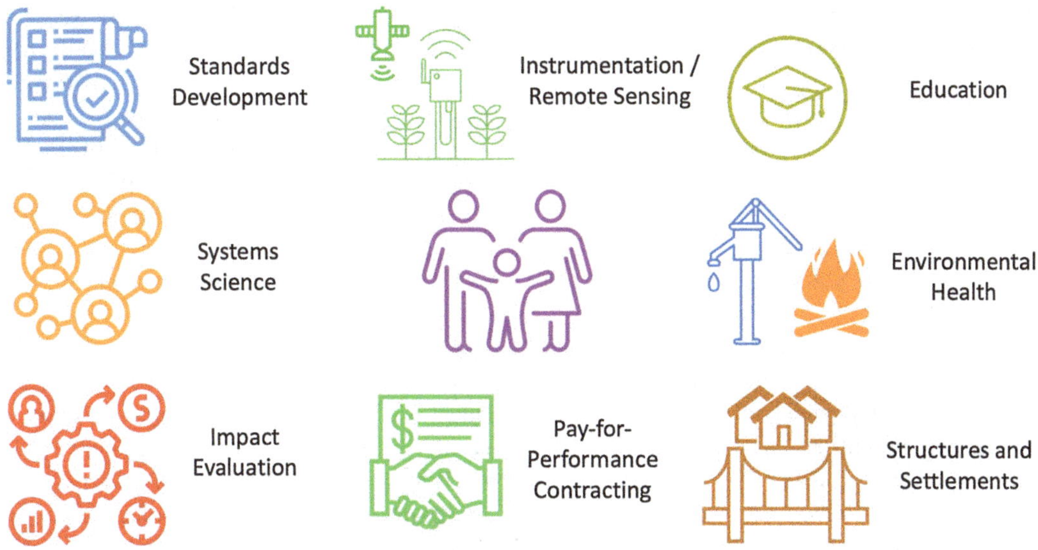

Fig. 1.7 Examples of Global Engineering education, research and practice

applications (Fig. 1.7), including those supported by the Mortenson Center in Global Engineering at the University of Colorado Boulder.

1.5.1 Shelters and Settlements

Engineering and design contributions to improved shelter and settlement design in global development have shifted in recent years from providing emergency and temporary assistance toward a more comprehensive, settlement-based approach. This shift corresponds to the increasing overlap in humanitarian assistance and longer term development.

Typically, humanitarian and disaster relief efforts are deployed separately from conventional development efforts. Humanitarian relief usually occurs in response to natural disasters, conflicts, and displacements. Engineering in this context has focused on rapid response in camps for refugees and other displaced people, with the general intent to provide services on a provisional basis.

However, the humanitarian relief and global engineering and development sectors are increasingly overlapping and becoming more integrated. The drivers that displace people are often linked to chronic development challenges at home, including food and water insecurity and climate change. Some observers have described the engineering-related intersection of these sectors *as "Peace Engineering,"* (Amadei 2019).

The United States Agency for International Development (USAID) Office of Foreign Disaster Assistance has a dedicated "humanitarian shelter and settlements" program that works to ensure safe and appropriate housing for disaster-affected communities, including considerations of culturally appropriate design, privacy, security, water and sanitation services, and future disaster risk reduction measures. This emerging application of global engineering "bridges the humanitarian–development nexus, from pre-disaster resiliency to emergency and temporary shelter, to the long-term development of sustainable housing and settlements," (Javernick-Will 2020).

Examples of such programs include Build Change (www.buildchange.org), which works to avert the worst impacts of natural disasters through constructing earthquake and typhoon-resistant housing; while architects at Mass Design Group (massdesigngroup.org) apply high-end architectural design in settings that have typically not benefited from these skills, including the creation of a Partners in Health hospital in rural Rwanda. Meanwhile, the Revitalizing

Informal Settlements and their Environments Program (RISE, www.rise-program.org) integrates wetlands management with water and sanitation services to develop healthier environments for groups of informal settlements in Indonesia and Fiji.

1.5.2 Remote Sensing

Space-based Earth observation instruments, while often funded, designed, and operated to serve the particular interests of wealthy countries, can provide benefits to developing countries at minimal additional cost. The insights gained from the analysis of remotely sensed data can result in practical actions as well as informing policy and public response.

A compelling example has been demonstrated in east Africa. Average rainfall in the region has declined over recent decades, decreased by over 20% in some areas since 1990 (Tierney et al. 2013). As a result, millions of people who live in the arid, drought-prone regions of the East African Rift Valley, including parts of Ethiopia and Kenya, suffer from a lack of safe, reliable, and affordable water (Viste et al. 2013; Shiferaw et al. 2014). The 2011 drought in East Africa caused food shortages for over 10 million people and as many as 260,000 deaths (Shabelle 2011; Nicholson 2014). The more recent 2016 drought in Kenya resulted in over 3 million people facing food insecurity (Uhe et al. 2017). These recent drought conditions represent an acute threat, and highlight the urgency of environmental changes driving water shortages, resulting in both public health and security emergencies.

Doris Kaberia, a Kenyan expert in food security and pastoral livelihoods, and Petros Birhane, an Ethiopian agricultural engineer and disaster relief expert, have dedicated their careers to building resilience in these regions of Africa. Petros and Doris' share their stores in Chaps. 8 and 9 of *The Global Engineers*.

Doris and Petros have also worked with the USAID-founded Famine Early Warning Systems Network (FEWS NET), designed to combat the worst consequences of drought-driven food insecurity. The FEWS NET model is based in part on rainfall and crop health estimates using remote sensing data to forecast food security stress based on estimated agricultural yields (Brown 2008; Senay et al. 2014; FEWSNet 2019). FEWS NET publishes food insecurity forecasts leveraged by national governments and international relief agencies to position food relief before the most severe consequences are felt by local populations.

Beyond the immediate, technocratic benefits of FEWS NET findings, such food insecurity models are raising repeated alarms. However, forecasts predicting humanitarian crises of increasing severity and frequency may have the effect of drawing political and public attention to the unjust impacts of climate change.

The NASA- and USAID-funded SERVIR program similarly works to adapt satellite-based remote sensing to map data, produce hydrologic and agronomic models, and develop decision aids to assist regions of Africa, Southeast Asia, and the Himalayas. These tools have assisted with water resource management, flood prediction, and land use planning, while building the capacities of developing countries to develop and manage these technical services (Hardin et al. 2005; Irwin et al. 2008; Wang et al. 2011; Al-Hamdan et al. 2017).

1.5.3 Instrumentation

Many development programs, from household-scale interventions to large-scale infrastructure, rely on third-party funders and lengthy processes for proposal development, implementation, and some measure of monitoring and evaluation. Despite increasing emphasis on the monitoring and evaluation phase, however, the reality of finite and time-bound funding often means that donors do not receive information about medium- and long-term impacts within developing countries which could inform their funding decisions.

This information and knowledge asymmetry can in part be addressed through improvements in the technologies used to collect ongoing data on the performance of interventions and the services delivered (Thomas 2016). Technological innovations in the design, deployment, and validation of instrumentation, and furthermore the analysis of data generated, can be used to inform programs, policies, and donors.

For example, a set of ongoing activities in Ethiopia and Kenya aims at improving the functionality of rural water supplies, reducing the downtime between repair activities, and ultimately improving water services, reducing water insecurity, and reducing the impacts of drought. The intervention includes installing satellite and cellular-connected sensors monitoring the runtime of rural electric pumps and linking these data through algorithms to online dashboards. The data are intended to be used by regional maintenance providers, utilities, national government entities, and international donors to both enable and support increased prioritization of repair services (Thomas et al. 2019).

Chantal Irbagiza, a Rwandese engineer, helped pioneer these approaches when she became frustrated by the regular breakdowns of rural handpumps in Rwanda. Chantal's career path is shared in Chap. 5.

While instrumentation represents a technical intervention, the insights generated are intended to inform policymakers, local and national budgeting, and donor decisions in an effort to recognize and address the major gap between the funding available for infrastructure installation and the funding available for operation and maintenance, both at the scale of the programs themselves, and by informing discussions at a global level.

This effort requires the participation of engineers from a broad range of disciplines, including civil, environmental, mechanical, and electrical engineering and computer science, integrated with expertise in governance, foreign aid, and community strengthening. A series of evaluations, including an independent impact evaluation, are currently being conducted to study these efforts. The instrumentation used has the unusual status of being both part of the intervention's theory of change, while also serving as the primary data collection tool used to evaluate the effectiveness of the intervention activities.

1.5.4 Impact Evaluation

The past decade has seen an increasing emphasis on the use of rigorously designed, independent experiments to evaluate development interventions. The academic fields of Global Health and Development Economics have been prolific in designing and administering rigorously designed experiments. Books such as *Poor Economics*, by Abhijit V. Banerjee and Esther Duflo, recipients of the 2019 Nobel Prize in Economics, review the sometimes counterintuitive results of these studies, which have catalyzed discussions among policymakers, national governments, donors, implementers, and researchers, and have attracted increasing participation from the public.

The results of these trials inform international policies, donor decisions, and implementation designs, and have leveraged best practices in academic research to build a body of knowledge over time. For example, a series of rigorous efficacy trials of chlorine interventions, including a multimillion-dollar water and sanitation trial, funded by the Gates Foundation, recently reported no effect on diarrhea among children under the age of 2 (Luby et al. 2018; Null et al. 2018). Meanwhile, other recent trials have shown considerable impact on diarrhea among children aged under 5 associated with household water treatment (Barstow et al. 2018). While sometimes providing contradictory results, the outcome is overall growth in the body of knowledge and consensus over time regarding the most effective and appropriate health interventions.

Limitations of the most rigorously designed evaluations and randomized controlled trials (RCTs) include the often small-scale of the studies, the constraints in adjusting the intervention during the study, and challenges in generalizability. Furthermore, the results of these studies often take many years to be analyzed and published.

Global Engineering may have a role in advancing impact evaluation methods that may be more adaptable to programs, more scalable, and more quickly actionable. Engineers do not normally conduct RCTs in the design, testing, and validation of new technologies. Instead, they design standardized and customized testing routines, gather data on technology performance, conduct analysis, and build in safety factors. This approach, which is iterative and exploratory, is no less rigorous, as evidenced by the absence of RCTs supporting or refuting the effectiveness of parachutes (Smith and Pell 2004). Engineers can adapt these approaches to evaluate global development interventions, especially those leveraging technological interventions. Study designs that allow for iterative implementation may still collect credible data. This approach is aligned with Implementation Science, an emerging field in medicine and public health. Implementation Science works to close the "know-do" gap between the demonstrated efficacy of a given health intervention and the actual observed effectiveness when deployed operationally (Fixsen et al. 2015).

1.5.5 Standards Development

Engineering has a rich history of developing, validating, refining, and implementing standards. These standards evolve based on evidence, best practice, and consensus building. Published standards can support the objective evaluation of products and services. Within the global development space, engineers have contributed to standards for household drinking water products (World Health Organization 2011), household cookstoves (International Organization for Standardization 2018), emergency shelters (UNHCR 2006), and other products and services.

The development and support of standards represents a valuable contribution by engineers to global development, and can be further applied to other areas of sanitation, energy, water, transportation, and infrastructure.

1.5.6 Pay for Performance

There is emerging alignment between rigorous performance measures and funding incentives. Such approaches can increase the accountability and scale of effective interventions (Sedlmayr 2019). Sometimes these are called performance-based payments, or formed as development impact bonds (DIBs).

One of the first DIBs in Africa was supported by USAID, the United Kingdom Department for International Development, the fund manager Instiglio, and the implementer Village Enterprise. The DIB is designed to facilitate the launching of small businesses in Uganda and Kenya, wherein "Village Enterprise gets up-front funding in the form of working capital from socially-motivated investors, and flexibility to adapt the program to maximize impact. USAID and other funders' repayments to the investors are conditional on Village Enterprise delivering verifiable results such as improved income and consumption," (https://www.usaid.gov/GlobalDevLab/innovation/village-enterprise-dib).

Engineers have helped design many of these performance-based contracting approaches. For example, as part of a team, engineers designed the first carbon credit-financed household water treatment programs, implemented in Kenya and Rwanda. These programs required private funding for implementation, followed by monitoring and issuance of carbon credits tied to ongoing performance. The credits were sold to commercial and concessionary buyers including the World Bank (Thomas 2016).

Jean Ntzinda, a Rwandese development professional, facilitated the introduction of these innovative funding mechanisms in Rwanda. Jean's story is told in Chap. 6 of this book.

Recently, iDE, a US-based technology and social enterprise innovator, launched the first development impact bond for sanitation (https://www.ideglobal.org/press/cambodia-rural-sanitation-dib). The $10 million-dollar fund is sponsored by the Stone Family Foundation, who provide funding to iDE to deploy sanitation interventions. Verified performance is then rewarded by USAID with payments back to the Stone Foundation.

Similarly, Bridges to Prosperity (B2P), a US-based nonprofit that designs and constructs pedestrian footbridges in developing countries, and is led by civil and structural engineers, has begun to transition from a reliance on charitable donations for one-off projects, to a scalable, accountable, and financially sustainable outcome-based model.

Isolation caused by lack of transportation infrastructure affects almost every facet of life for the rural poor. Without adequate transportation, families cannot access schools, health care, employment, or local markets to sell and buy goods. The World Bank estimates that nearly a billion people worldwide lack access to an all-season road within 2 km, illustrating the scope of the problem and the challenge of addressing it at scale. B2P has constructed more than 300 footbridges in 20 countries, an infrastructure intervention that is cost-effective, durable, and relatively simple to scale. An economic impact evaluation of B2P's footbridges in Nicaragua found a 35.8% increase in labor market income attributable to the access provided by the bridges (Brooks and Donovan 2019).

B2P's field program in Rwanda started in 2012 and has led to the completion of 47 footbridges that have created safe access for an estimated 274,000 people. Over the next five years, B2P plans to construct approximately 350 footbridges in the country. This rapid program growth presents an unprecedented opportunity for rigorous investigation of the effects of new footbridges on a number of key economic, health, agricultural, and education outcomes for rural communities.

B2P's scale-up model is designed to combine funding from local and national governments in Rwanda with debt financing from international sources. Upon completion of the footbridges, and demonstration of the health and economic impacts, an outcome-based payment will be issued to B2P to repay investors and further expand their program.

The CEO of Bridges to Prosperity, Avery Bang, is a civil engineer dedicated to mobilizing capital ethically, toward helping to solve rural isolation as a barrier to economic prosperity. Avery is profiled in Chap. 7 of *The Global Engineers.*

1.5.7 Systems Engineering and Science

Engineers have begun to recognized the importance of considering the complex systems governing effective development when designing projects. In this regard, engineers have promoted "systems thinking," systems engineering, and the inclusion of governance and institutions in addressing basic service delivery (Amadei 2015).

Systems engineering has been particularly leveraged in the fields of water and sanitation (Walters and Javernick-Will 2015; Davis et al. 2019). For example, USAID promotes a local systems framework for the development of water supplies, the design of which includes principles such as systems mapping, holistic design, monitoring, and accountability (USAID 2014). The USAID Sustainable WASH Systems (SWS) activity, led by the University of Colorado Boulder, is designed to characterize the systems behind water, sanitation and hygiene (WASH) services across several programs in Ethiopia, Kenya, and Uganda. The activity recognizes past failures in providing reliable service delivery and seeks to generate knowledge around the broader systems required to improve services. One element of the SWS model is the promotion and facilitation of learning alliances, which bring together actors at the district and local level to collaborate in identifying relevant service delivery relationships, material flows, and leverage points, and through collective action improve WASH services (USAID 2018).

Dan Hollander, an American civil engineer and former USAID Foreign Service Officer, leads the SWS program. Dan's experience working in low-, middle-, and high-income countries as an engineer is shared in Chap. 10.

Similarly, the WASH Agenda for Change, supported by a network of influential donors and implementers, considers financial planning as an underpinning activity in promoting improved WASH services. The model also promotes collective action and measurement of long-term service delivery (Agenda for Change 2019).

A potential limitation of systems approaches is the continued focus on local actors and factors, while the broader implications of globalization are considered outside of the design envelope. Systems engineering would benefit from considering global trade imbalances, resource exploitation, and the unequal distribution of wealth, which can preclude local governments from considering tax-based resources as relevant factors when exploring the relative influence of different factors in the sustainability of basic services.

1.5.8 Education and Training

There are several emerging educational programs aligned with the motivations of Global Engineering. The Engineering for Change (E4C https://www.engineeringforchange.org) curriculum, offered online, was designed and curated by the American Society of Mechanical Engineers (ASME), with participation from the Institute of Electrical and Electronics Engineers (IEEE) and Engineers Without Borders-USA. The free online curriculum offers introductory training sessions for engineers seeking to apply their skills to global development challenges, including introductions to development history, practice, and local contexts.

Iana Aranda, a mechanical engineer and the president of Engineering for Change, is dedicated to developing the Global Engineering workforce of the future. "We face a deficit in the global engineering workforce that could escalate into a crisis if we do not act."

According to a World Economic Forum report in 2015, among all engineering jobs, it is estimated that there is only one qualified engineer available for every 1.9 positions (WEF 2015). "This shortage of engineers grows more acute daily, particularly in frontier markets. One strategy in addressing this is to encourage young people to enter technical fields, and to understand why engineering matters and how they can contribute. To reach young, potential engineers, traditional training must be supplemented with new approaches to engaging, informing and deploying this critical technical workforce. Digital communities and knowledge platforms provide an accessible pathway."

As part of the solution, Iana launched the E4C Fellowship, a "workforce development program in social innovation that serves to build engineering capacity and prepare talent to solve local and global challenges." E4C Fellows are young engineers recruited from around the world conducting project and product research to support the E4C Solutions Library, while building their own professional networks and Global Engineering capacity.

Other organizations worldwide also offer service learning, volunteer, and training opportunities for engineering students and professionals. Many of these organizations focus on small-scale, village-level partnerships, while others also offer professional consulting services for larger scale efforts. These programs include Engineers Without Borders-USA (www.ewb-usa.org), many of the EWB organizations affiliated with EWB-International (www.ewb-international.com/countries), Engineers in Action (www.engineersinaction.org), and Engineers for a Sustainable World (www.eswglobal.org).

At the university level, the Centre for Global Engineering at the University of Toronto is a cross-disciplinary research institute that focuses on areas of global need, including food and nutrition, water and sanitation, health and shelter. The Centre involves the participation of all engineering disciplines in the Faculty of Applied Science and Engineering, and works in both Canada and developing countries.

Similarly, the Mortenson Center in Global Engineering at the University of Colorado Boulder has evolved a model of scale-appropriate technology design and implementation, with an increasing emphasis on the development and validation of more broadly applicable methods, technologies, and evidence generation. As reflected in the name change from "Engineering for Developing Communities" to "Global Engineering," the Center seeks to positively impact vulnerable people and their environment by improving development tools and practice.

Areas of research at the Center include organizational theory and systems engineering, the development and validation of water, sanitation, energy, infrastructure and agricultural technologies and methods, design of service delivery models, impact measurement methods and technologies including instrumentation and remote sensing, and the development of standards for engineered systems applied in disaster relief. The Mortenson Center curriculum includes opportunities to take short courses in complementary topics such as Global Health, Development Economics, remote sensing, statistical analysis, and impact evaluation. The required field practicum embeds students within global development agencies for at least three months, with some students continuing to engage with these agencies for many years.

1.6 This Book

Discussions among students, faculty, staff and partners within the Mortenson Center in Global Engineering at the University of Colorado Boulder revolves around several key questions. What is our role as global citizens and as engineers? What challenges are we trying to solve, and how? How can we participate, without unwittingly and unwillingly engaging in neo-colonialism? Are we reinforcing autocracies and unaccountable institutions? Or, are we incrementally but meaningfully contributing to a more just and equitable world?

These are hard questions for anyone to answer —including college students who are only recently becoming aware of their own privileges, and feel passionate about helping to create a more just world.

Engineers and students have often sought a textbook solution to global development challenges. Occasionally, such books are offered, but unfortunately global development is uncertain, complex, and evolving. Often the dogma of "right" and "wrong" approaches is impervious to evidence and context, and instead reflects institutional incentives and motives.

This book seeks to examine the role and ultimately the impact of engineers in global development. It presents case studies of programs and technologies with which the author has been personally involved, and profiles engineers and other aligned professionals working within these and related programs.

Together, engineers can work alongside countries, communities, and other professionals to identify and dismantle the underlying causes of persistent global poverty, and elevate all people, and their environment, as Global Engineers.

References

Agenda for Change (2019) Agenda for Change

Al-Hamdan MZ et al (2017) Evaluating land cover changes in Eastern and Southern Africa from 2000 to 2010 using validated Landsat and MODIS data. Int J Appl Earth Obs Geoinf. https://doi.org/10.1016/j.jag.2017.04.007

Amadei B (2014) Engineering for sustainable human development: a guide to successful small-scale community projects. ASCE

Amadei B (2015) A systems approach to modeling community development projects. Momentum Press

Amadei B (2019) Engineering for peace and diplomacy. Sustainability (Switzerland). https://doi.org/10.3390/su11205646

Amadei B, Sandekian R, Thomas E (2009) A model for sustainable humanitarian engineering projects. Sustainability. https://doi.org/10.3390/su1041087

Amadei B, Thomas E (eds) (2020) UNESCO Engineering Report—Engineering the SDGs. Paris

Barstow C et al (2018) Health, livelihood, and environmental impacts of the distribution of a carbon-credit-financed, large-scale water filter and improved cookstove programme in Rwanda. Lancet Planetary Health. https://doi.org/10.1016/S2542-5196(18)30116-5

Bourn D, Neal I (2008) The global engineer incorporating global skills within UK higher education', *Dfid*

Brooks W, Donovan K (2019) Eliminating uncertainty in market access: evidence from new bridges in Rural Nicaragua. Econometrica (in revision)

Brown ME (2008) Famine early warning systems and remote sensing data, Famine Early warning systems and remote sensing data. https://doi.org/10.1007/978-3-540-75369-8

Centre for Applied Research, N. S. of E. et al (2015) Financial flows and tax havens combining to limit the lives of billions of people, (December), p 113

Davis A, Javernick-Will A, Cook S (2019) The use of qualitative comparative analysis to identify pathways to successful and failed sanitation systems. Sci Total Environ 663:507–517. Elsevier B.V. https://doi.org/10.1016/j.scitotenv.2019.01.291

FEWSNet (2019) FEWS NET data center, food security classification data. http://fews.net/fews-data/333

Fixsen D et al (2015) Implementation science. In: International encyclopedia of the social & behavioral sciences, Second edn. https://doi.org/10.1016/b978-0-08-097086-8.10548-3

World Economic Forum (2015) The human capital index

Haines A, Ebi K (2019) The imperative for climate action to protect health. New Engl J Med. https://doi.org/10.1056/NEJMra1807873

Hardin D et al (2005) Visualizing earth science data for environmental monitoring and decision support in Mesoamerica: the SERVIR project, AGU Spring Meeting Abstracts

Institute for Health Metrics and Evaluation (2019) Global burden of disease compare tool. http://vizhub.healthdata.org/gbd-compare

International Organization for Standardization (2018) ISO/TR 19867-3:2018—Clean cookstoves and clean cooking solutions—Harmonized laboratory test protocols—Part 3: Voluntary performance targets for cookstoves based on laboratory testing

Irwin D et al (2008) SERVIR: a regional monitoring and decision support system for Mesoamerica, AGU Spring Meeting Abstracts

Javernick-Will A (2020) Special issue 'sheltering and housing displaced populations'. https://www.mdpi.com/journal/sustainability/special_issues/Sheltering_Housing_Displaced_Populations. Accessed 3 Jan 2020

Luby SP et al (2018) Effects of water quality, sanitation, handwashing, and nutritional interventions on diarrhoea and child growth in rural Bangladesh: a cluster randomised controlled trial. Lancet Glob Health. https://doi.org/10.1016/S2214-109X(17)30490-4

Mintz K et al (2014) Integrating sustainable development into a service-learning engineering course. J Prof Issues Eng Educ Pract. https://doi.org/10.1061/(ASCE)EI.1943-5541.0000169

Mitcham C, Munoz D (2010) Humanitarian engineering. Synthesis Lectures on Engineers, Technology and Society. https://doi.org/10.2200/s00248ed1v01y201006ets012

Nicholson SE (2014) A detailed look at the recent drought situation in the Greater Horn of Africa. J Arid Environ. https://doi.org/10.1016/j.jaridenv.2013.12.003

Nilsson L, Madon T, Sastry SS (2014) Toward a new field of development engineering: linking technology design to the demands of the poor. In: Procedia engineering. https://doi.org/10.1016/j.proeng.2014.07.032

Null C et al (2018) Effects of water quality, sanitation, handwashing, and nutritional interventions on diarrhoea and child growth in rural Kenya: a cluster-randomised controlled trial. Lancet Global Health. https://doi.org/10.1016/S2214-109X(18)30005-6

OXFAM (2017) [C2] An Economy for the 99%, OXFAM. ISBN 978-1-78077-993-5

Peet R (2014) The tyranny of experts: economists, dictators, and the forgotten rights of the poor. J Econ Geogr. https://doi.org/10.1093/jeg/lbu022

Robertson A (2018) OLPC'S $100 laptop was going to change the world—then it all went wrong, The Verge. https://www.theverge.com/2018/4/16/17233946/olpcs-100-laptop-education-where-is-it-now. Accessed 3 Jan 2020

Rosling H, Rosling O, Rönnlund AR (2018) Factfulness: ten reasons we're wrong about the world–and why things are better than you think. Flatiron Books, New York

Sedlmayr R (2019) Rewarding poverty alleviation: a case study in payment-by-results. Working Paper

Senay GB et al (2014) Drought monitoring and assessment: remote sensing and modeling approaches for the famine early warning systems network. Remote Sensing and modeling approaches for the famine early warning systems network. In: Hydro-meteorological hazards, risks, and disasters. https://doi.org/10.1016/b978-0-12-394846-5.00009-6

Shabelle L (2011) Drought-related food insecurity: a focus on the Horn of Africa. Drought Emergency

Shiferaw B et al (2014) Managing vulnerability to drought and enhancing livelihood resilience in sub-Saharan Africa: technological, institutional and policy options. Weather Clim Extrem. https://doi.org/10.1016/j.wace.2014.04.004

Smith GC, Pell JP (2004) Parachute use to prevent death and major trauma related to gravitational challenge: Systematic review of [randomized] controlled trials. J Int Assoc Phys AIDS Care. https://doi.org/10.1080/20954816.2017.1310791

Stellar D (2010) The playpump: what went wrong? Earth Institute. https://blogs.ei.columbia.edu/2010/07/01/the-playpump-what-went-wrong/. Accessed 3 Jan 2020

The World Bank (2018) Nearly half the world lives on less than $5.50 a day. https://www.worldbank.org/en/news/press-release/2018/10/17/nearly-half-the-world-lives-on-less-than-550-a-day

Thomas E (2019) Toward a new field of global engineering. Sustainability 11(14): 3789. Multidisciplinary Digital Publishing Institute. https://doi.org/10.3390/su11143789

Thomas EA (2016) Broken pumps and promises: incentivizing impact in environmental health. Springer

Thomas EA et al (2019) Quantifying increased groundwater demand from prolonged drought in the East African Rift Valley. Sci Total Environ. https://doi.org/10.1016/j.scitotenv.2019.02.206

Thomas E, Amadei B (2010) Accounting for human behavior, local conditions and organizational constraints in humanitarian development models. Environ Dev Sustain. https://doi.org/10.1007/s10668-009-9196-1

Tierney JE et al (2013) Multidecadal variability in East African hydroclimate controlled by the Indian Ocean. Nature. https://doi.org/10.1038/nature11785

Uhe P et al (2017) Attributing drivers of the 2016 Kenyan drought. Int J Climatol 38(Suppl 1):e554–e568. https://doi.org/10.1002/joc.5389

UNFCCC (2020) The list of least developed countries. https://unctad.org/en/Pages/ALDC/LeastDevelopedCountries/UN-list-of-Least-Developed-Countries.aspx. Accessed 3 Jan 2020

UNHCR (2006) Emergency Shelter Standards, (June), pp 1–21

UNICEF (2018) Levels & trend in child mortality: estimates developed by the estimates developed by the UN Inter-agency Group for UN Inter-agency Group for Child Mortality Estimation Child Mortality Estimation

USAID (2014) Local systems: a framework for supporting sustained development. https://www.usaid.gov/policy/local-systems-framework

USAID (2018) Finding new solutions to strengthen local systems and improve WASH service delivery and sustainability

Viste E, Korecha D, Sorteberg A (2013) Recent drought and precipitation tendencies in Ethiopia. Theoret Appl Climatol. https://doi.org/10.1007/s00704-012-0746-3

Walters JP, Javernick-Will AN (2015) Long-term functionality of rural water services in developing countries: a system dynamics approach to understanding the dynamic interaction of factors. Environ Sci Technol. https://doi.org/10.1021/es505975h

Wang J et al (2011) The coupled routing and excess storage (CREST) distributed hydrological model. Hydrol Sci J. https://doi.org/10.1080/02626667.2010.543087

Woodward D (2015) Incrementum ad absurdum: global growth, inequality and poverty eradication in a carbon-constrained world. World Econ Rev

World Bank (2019) Net official development assistance and official aid received (current US$). https://data.worldbank.org/indicator/DT.ODA.ALLD.CD

World Health Organization (2011) Evaluating household water treatment options: Heath-based targets and microbiological performance specifications. WHO Press. https://doi.org/10.1016/j.ijrobp.2007.10.057

World Health Organization (2018) COP24 special report: health and climate change, p 73. https://apps.who.int/iris/bitstream/handle/10665/276405/9789241514972-eng.pdf?ua=1

An Engineer's Education

2

Abstract

This chapter presents the career path of one global engineer, growing from volunteer international travel to scaling a nationwide health program in Rwanda over 16 years.

2.1 Introduction

Often, my job looks like I am simply a professional emailer. My day hasn't typically included jumping off 10-meter platforms into wave pools, or being flipped over in a sinking helicopter simulator in the dark, or fighting blinding fires and freezing floods. Or, perhaps even harder, sewing cuffs or moving marbles from one table to another with Dixie cups and rubber bands. But at least I nailed the Etch-a-Sketch challenge.

In 2017, I was a finalist for the Canadian Astronaut Selection process. I was challenged and tested along with nearly 4,000 other applicants. Finally, the CSA got wise, and now my friends Josh and Jenni are currently training in Houston as astronauts to work on the International Space Station (ISS).

The International Space Station (ISS) has been inhabited continually since November 2000. Astronauts from 18 countries have conducted cutting-edge research, learnt how to live for long durations in space, and inspired people around the world. Indeed, the ISS motto is "Off the Earth, for the Earth."

Keeping astronauts alive and healthy in space requires access to clean water and air, both of which are continuously recycled or scrubbed on the ISS. However, while we have astronauts orbiting the planet, over a billion people on Earth still do not have clean water; even more lack safe sanitation and billions of people cook with unclean fuel every day. These environmental hazards constitute some of the leading causes of illness and death globally. Ninety-nine percent of children under the age of 5 who die every year live in developing countries—and many of those children become sick due to exposure to dirty water and dirty air.

The Constellation program, led by former astronauts Ron Garan, Nicole Stott, Leland Melvin, and Anousheh Ansari, have embraced this dichotomy as a call to action. "As astronauts, we were each deeply transformed by our experiences in space—both by what we learned and by what we achieved by working with others so different than ourselves. We believe in our hearts that humankind and the Earth are highly interdependent by nature. All human actions are interconnected with one another and with our planet. Profound collaboration is the key to making what seems impossible possible."

E. Thomas, *The Global Engineers*, Sustainable Development Goals Series, https://doi.org/10.1007/978-3-030-50263-8_2

2.2 Getting a Global Education

My first experience of developing countries occurred in high school as a student journalist with my high school newspaper. In 2000–01, we visited Cuba followed by Vietnam, two countries with long histories of American interference and attempted subjugation. I heard propaganda from the US Interests Section in Havana, and marched in a protest demanding the return of the young boy Elian Gonzalez, whose mother had drowned trying to get him to the United States. I met the US Ambassador to Vietnam, who was shot down and imprisoned alongside John McCain during the Vietnam War, and later declared that the Vietnamese Communism had never been a threat to the United States.

These experiences gave me a sense of the complexity of our world, and a drive to travel productively—not as a tourist. In 2001, during my freshman year at the University of Colorado Boulder, I met Professor Bernard Amadei, who had just founded Engineers Without Borders-USA. In 2003, I traveled as a member of this organization to Nepal, where I installed a few computers in classrooms in Himalayan villages. On my return, I wrote an article for the *Colorado Engineer Magazine* describing work being done by the Himalayan Light Foundation, a nonprofit organization, to reduce brain drain from rural villages.

I first traveled to Rwanda in 2004 at the invitation of an American nurse posted at the US Embassy who had observed poor health outcomes potentially associated with contaminated drinking water in a community in the west of the country. Over several years, our volunteer team worked to fundraise, design, and install rainwater catchment and water treatment systems in several communities. Although the project received positive press at home, the long-term functionality and benefits of these systems were much more elusive.

So why Rwanda? Almost 25 years ago, Rwanda suffered a genocide that killed nearly a million people. Today, Rwanda is considered among the least corrupt and most progressive countries in Africa. It has one of the rapid GDP growth rates in the region, and the fastest annual decline in child deaths globally. However, pneumonia and diarrhea still remain the leading causes of illness and death among children.

One of our first volunteer projects involved installing solar panels on a hospital in western Rwanda, and providing emergency lighting to the operating rooms. Our technical training for the doctor was interrupted, however, by a summons to return to the hospital. The doctor beckoned me to come along, and soon seemed to be in a heated argument with one of his staff. We went to one of the dimly lit wards, which smelled like ammonia and sweat. A young woman was sitting on a bed laughing and looking around, while a small baby lay in a crib beside her. The baby was receiving oxygen and seemed to be sleeping. The doctor checked on the baby, increased the oxygen level, and sat down on the bed.

The young woman who had given birth the previous week suffered from mental health issues. We were told that she was a survivor of the 1994 genocide and the victim of frequent rape. Once the doctor had delivered her baby and sent her home, she appeared to have forgotten its existence. Neighbors found the newborn child, malnourished, and alerted the doctor, who had to make a decision—whether to allow the baby to die in agony in a village, or ease its suffering and let it die painlessly in the hospital. There were no orphanages to take the child, and nobody else in the community stepped forward to adopt. At this moment, the staggering realities of global inequalities displaced the happier story of productive voluntourism.

In 2004, I began working as a civil servant at NASA in Houston, receiving attention from recruiters in part thanks to articles I had written about our work in Rwanda. At NASA, I established the Engineers Without Borders-Johnson Space Center Chapter, which attracted engineers, astronauts, mission controllers, and educators to our lunch meetings and weekend fieldwork events.

My work at NASA centered on the Life Support and Habitability Systems Branch. As an aerospace engineer, I designed and managed technologies for water management, water quality detection, water recycling, and air quality detection. Some of these systems formed part of the Space Shuttle and the International Space Station.

During this time, I had two parallel lives—trying to advance a career in NASA, and trying to be faithful to the commitments we had made in Rwanda. As a volunteer organization, we struggled with the quality and commitment of our volunteers, the availability of funding, and the productivity of our group. Beyond these functional concerns, I became increasingly concerned that project-level efforts, despite a focus on local ownership and low-tech interventions, simply were not sustainable in the most basic sense—they failed to continue working. We needed a way to tie funding to actual services—and not to one-off projects.

2.3 Broken Pumps and Promises

This emphasis on discrete projects is not limited to small organizations. Driving along rural roads in many developing countries, you see frequent evidence of the generous intentions of global humanitarian aid agencies. A tangible example is the hand-driven water pumps that dot the landscape, intended to provide clean water to more people. Thousands of these pumps, funded and implemented by organizations large and small, are installed every year in developing countries.

However, you can never predict whether the next water pump you pass will be surrounded by people, often women and children, filling their jerry cans, or will stand as a decrepit artifact of wasted resources.

Studies show that between 30 and 80% of water pumps fail within a year of installation (Foster 2013). While the proximate failures may be a leaky seal, a broken riser, or a missing handle, these are only symptoms of a wider failure in terms of how we fund, incentivize, and monitor these efforts.

Some experts suggest that the cost of installing a new hand pump, is equivalent to funding the operation and maintenance of its neighbor for a century. Or, put another way, an implementer with 500 installed pumps, could choose either to install 100 new hand pumps in a year or to maintain the original 500 for 20 years.

However, this choice is dependent on the ability to persuade funders that maintenance as interesting as new construction. At present, funders continue to focus on construction, and sustainability is usually addressed through "participatory community development," where local communities are, in theory, empowered to manage their own water supplies. In reality, this approach has often failed to produce cost-effective interventions. Moreover, these challenges exist not just for hand pumps, but also for myriad health and environmental interventions both in developing and developed countries.

Traditional development organizations rely on government, UN, or philanthropic grants, and have finite funding tied to discrete projects targeting specific goals.

Typical funders, ranging from church and community groups, and universities, all the way up to the UK Department for International Development, the US Agency for International Development, and the World Bank, provide funding for projects that are intended to improve the health and livelihood of people in developing communities. These include water pumps and water filters, cookstoves, latrines, and solar lighting systems.

Such funding usually lasts for a couple of years. During that time the implementers will try to evaluate the project's impact. If there are sufficient funds, this might include running a randomized controlled trial to see if the projects are improving health or other outcomes. However, eventually the funding runs out, and everyone moves on.

In recent years, attempts to prioritize long-term support for these kinds of interventions have emerged. These include pay-for-performance contracting mechanisms (sometimes in the form of output-based aid or impact bonds), increasing emphasis on institutional systems and governance, and evidence generation.

2.4 Carbon Finance—A Novice's Guide

Unsafe drinking water and household air pollution are major causes of mortality worldwide. Over a billion people lack access to safe drinking water, more than a third of whom rely primarily on open wells and untreated surface water that can be contaminated with human and animal feces.

Cooking indoors on traditional open-fire stoves with solid biomass fuels such as wood and charcoal has been linked to pneumonia, low birth weight, and impaired development in children. Household air pollution is also associated with pulmonary and cardiovascular disease in adults. More than 80% of Rwandans rely on firewood as their primary fuel source (Kirby et al. 2019).

In Rwanda, most rural villagers drink untreated water and burn firewood on open stoves. Many efforts to address these environmental health issues have taken place—and continue to do so—often without the financial support to ensure long-term sustainability.

In 2007, an article published in the *New York Times*, about the United Nations carbon credit market, triggered an idea—could we hijack this system to fund ongoing support of water treatment in developing communities?

International carbon credit markets are designed to encourage sustainable, clean development around the world, while reducing current and projected emissions. The United Nations Clean Development Mechanism (UN CDM) has created a worldwide market for carbon credits from clean projects. However, few of those programs at the time were active in truly developing countries—with less than 2% of the projects registered in Africa.

In 2010, I co-founded (with NASA Astronaut Col. Ron Garan, among others) a social enterprise called Manna Energy Limited, the first organization to successfully register a UN CDM program combining drinking water treatment with carbon financing, and earn UN carbon credits for treating drinking water in rural Rwanda. The premise of the carbon credits addresses a prevailing practice among rural residents of boiling water with nonrenewable wood. When we install a water treatment system that treats water to WHO standards for microbiological contamination, we provide a clean alternative to these baseline practices.

Many people in Rwanda and elsewhere boil their water. Some use alternatives such as chlorination or solar distillation, while many consume untreated water and often suffer the health consequences.

Our first program in Rwanda only claimed carbon credits for about 5% of the water we treated—this represents a fraction of the water that was previously boiled (1–2 L per person per day) by a fraction of the residents that engaged in this practice (roughly 25%). As such, the program was not financially sustainable from the carbon credits, and therefore not scalable. This gap highlights some of the challenges inherent in encouraging clean development in regions where the baseline energy use is below global averages.

In an effort to address this disparity, the United Nations incorporated the concept of "suppressed demand," whereby emission reductions could be calculated based on projected future energy use. Suppressed demand involves estimating an energy demand or need that is not actually being realized and proactively making available a cleaner alternative. Carbon credits are then claimed on the "demand" for the baseline energy source, rather than status quo usage. This is controversial because the credits are given for emissions avoidance, rather than reductions of current emissions. In essence, suppressed demand can be thought of as prevention, rather than a cure.

The CDM Executive Board defines suppressed demand as a concept arising in circumstances where a "minimum service level that is able to meet basic human needs is unavailable to the end user of the service prior to the implementation of the project activity. This service level is a 'choice' that reflects that the service provided prior to the implementation of the project activity would increase if it were not suppressed by the lack of income and high unit costs of the service." The UN CDM incorporated this concept into a drinking water treatment methodology that allows project developers, including ourselves, to claim carbon credits for residents in regions where water boiling is otherwise a viable alternative.

Separate from the technical premise of suppressed demand for water treatment, which has been hotly debated, it is important to recognize the political considerations that led to this approach. With less than 2% of CDM programs active in Africa, the situation was clearly skewed in favor of developing economies rich enough to realize the tremendous demand for energy, while poorer countries, with the same demand for the health and welfare benefits from increased energy usage, were left out of the market.

The technical premise of suppressed demand is a tool to reach an important political end—increased engagement by the world's poorest countries in increased clean development. This approach has been criticized within the global development water sector as a misuse of carbon markets, since suppressed demand permits claims that exceed the current baseline use of fuel. Interestingly, these criticisms come primarily from the nongovernmental organization (NGO) water sector, rather than carbon credit technical experts.

Manna Energy Limited's first step was to work with the disease control textile company Vestergaard Frandsen to conceive and develop a carbon credit-financed water treatment program. The program was deployed in 2011 and has since provided water filters for over 4 million people in rural Kenya.

Ongoing monitoring of our carbon credit-financed projects is required by the registration authorities; carbon credits are not issued unless it can be demonstrated and independently verified that the water treatment systems are both functional and used by the target communities. This meets the definition of sustainability—the program continues to be financed when ongoing success is demonstrated. As a result, there is no longer a disconnect between funding and public health outcomes.

2.5 Tubeho Neza—Let Us Live Well

Later in 2011, we were hired by the UK-based water test kit company, DelAgua, to design a large-scale program, with private financing. As a product company with a track record of sales, DelAgua was in a strong position to secure such financing and demonstrate revenues through the generation and sale of UN carbon credits.

The program, called Tubeho Neza (meaning "let us live well"), was a partnership between DelAgua and the Rwandan Ministry of Health. In the fall of 2014, over 101,000 households with nearly half a million people from the poorest economic quartile of Rwanda's Western Province were selected to receive a Vestergaard Frandsen LifeStraw Family 2.0 table-top household water filter and an EcoZoom Dura high-efficiency portable wood-burning cookstove, together with community and household education and behavior change messaging. Each household was visited approximately every four months for a year following distribution.

In 2015, the program reached nearly 2 million more people with cookstoves. A staff numbering almost 1,000 worked across Rwanda, visiting nearly 350,000 homes. In cooperation with the Rwanda National Police and the Ministry of Health, we moved 470,000 products down muddy roads and into homes.

My colleague, Jean Ntazinda, helped lead this program and facilitated our partnership with the government of Rwanda. Jean's story is shared in Chap. 6.

The selection of these products was not without controversy. As described in the previous chapter, many professionals in the development sector emphasized the need to use locally produced products. However, the priority of the Ministry of Health was to achieve improved health outcomes, starting with the leading causes of illness among children: diarrhea and pneumonia. On this basis, we evaluated which technologies could most effectively reduce drinking water contamination and indoor air pollution. We also explored which technologies were most likely to be adopted by households, were eligible for carbon finance, and were durable and could be maintained by a team of technicians under our employ. Finally, we reached a consensus and decided to import water filters that exceed the WHO "highly protective" rating, and improved wood fuel cookstoves that would most easily be adopted by communities while still reducing emissions. In 2012, we conducted a pilot of 1,500 households to test the viability and impact of these products, and made subsequent adjustments to the products and our program design.

From 2012 to 2016, the London School of Hygiene and Tropical Medicine and Emory University, under the leadership of Professor Thomas Clasen, conducted a series of randomized controlled trials to evaluate our program. They studied its design; the adoption of the stoves and filters; our impacts on water quality, air quality, respiratory disease, and diarrhea; and critiqued the carbon credit financing mechanism.

The results were published in 2019 and showed that we had reduced the prevalence of reported diarrhea and acute respiratory infection in children under 5 years old by 29% and 25%, respectively (Kirby et al. 2019). These results suggest that programmatic delivery of household water filters and improved cookstoves can provide a scalable interim solution for rural populations that lack access to safe drinking water and rely on traditional fires for cooking. Over the same period of time, the results of several much larger studies (of much smaller interventions deployed for the purposes of research) indicated very poor impacts on health (Cumming et al. 2019). Our program contrasted strongly with these interventions, demonstrating the viability of bringing water filters and cookstoves to vulnerable households, and will thus help inform future national initiatives.

The reported respiratory-related health impacts were surprising. Other biomass-burning stove intervention studies found no effect on self-reported respiratory symptoms among cooks and/or children, leading to efforts to shift to cleaner fuels such as liquid petroleum gas (LPG).

Moreover, Clasen's team reported no reduction in personal exposure to fine particulate matter (PM2.5) from the stoves among adult cooks or children. They suggested that the impact on respiratory infection could be attributable in part to an immunological boost from experiencing fewer enteric infections.

Beyond the directly measured health impacts, we also analyzed the overall programmatic costs and benefits, including fuelwood savings, time savings, and environmental and health benefits. Over a five-year period, the total program cost is estimated at over \$11.91 million, while the total benefits were means valued at over \$66.67 million, for an estimated mean cost–benefit ratio of over 5.6 (Barstow et al. 2019). A sensitivity analysis of the major factors indicated a cost–benefit ratio range of approximately 1–16. The primary benefit identified was the environmental impact of wood fuel savings attributable to the improved cookstoves. The study estimated that 118,000 tones of annual wood fuel savings in the Western Province may be attributable to the program in year 1, decreasing to 65,000 tons in year 5.

These findings suggest that the program may help to compensate for the Government of Rwanda's projected regional wood fuel deficit of 106,000 tons per year by 2020.

Overall, the results suggest that the program was cost-effective in reducing wood fuel use, improving drinking water quality, and reducing the risks of diarrhea and respiratory illness among children under 5, pointing the way toward an interim solution for healthier living, while cleaner cooking solutions are developed and scaled to reach the poorest. As long as these products are still in use, DelAgua earns carbon credits which are then sold to the World Bank and other buyers. As of 2020, the water filters and stoves used in the trial are nearing the end of their lifetimes and it does not look as if they can be replaced, given the weak carbon credit market. As a result, DelAgua has transitioned to a focus on carbon-credit subsidized retail sales of these products in Rwanda. Over the same years as this program in Rwanda, there were several other large-scale trials of water, sanitation and hygiene interventions in low-income settings. They indicated little or no impacts on health. Similarly, a study of community hygiene clubs in Rwanda—designed to improve awareness without providing any products—showed no impact on health. Cleaner fuels, such as liquefied petroleum gas, may offer a potential solution to further improving air quality, and are being evaluated in a multi-country trial that includes Rwanda. But accessibility and affordability of such fuels will continue to be a challenge in many settings. In contrast, the Tubeho Neza programme has showed that it is possible to provide interventions against major diseases to vulnerable households at scale and to secure their adoption and consistent use. It also demonstrated the efficiency of combining critical environmental interventions at the household level. This achievement can inform other national efforts.

These efforts, in particular growing a program from a volunteer effort to a nationwide health intervention, have taught me about the role of engineers in working globally, and where we can best help address the underlying causes of persistent global poverty.

References

Barstow C et al (2019) A cost-benefit analysis of livelihood, environmental and health benefits of a large scale water filter and cookstove distribution in Rwanda. Dev Eng. https://doi.org/10.1016/j.deveng.2019.100043

Cumming O et al (2019) The implications of three major new trials for the effect of water, sanitation and hygiene on childhood diarrhea and stunting: a consensus statement. BMC Med. https://doi.org/10.1186/s12916-019-1410-x

Foster T (2013) Predictors of sustainability for community-managed handpumps in sub-Saharan Africa: evidence from Liberia, Sierra Leone, and Uganda. Environ Sci Technol 47(21):12037–12046. https://doi.org/10.1021/es402086n

Kirby MA et al (2019) Effects of a large-scale distribution of water filters on water quality and diarrhoea: a cluster randomized controlled trial in Western Province, Rwanda. PLOS Med 1–24. https://doi.org/10.1371/journal.pmed.1002812

3 Measuring Progress and Performance in Global Engineering

Abstract

Measurement and feedback are increasingly being incorporated into global engineering and development financing and programming. Within this ecosystem, there is room to incorporate near-time measurement of program, product, and service performance, and feed these measures back to responsible parties in a way that enables iterative testing and improvement. This chapter presents a history of feedback approaches, and reviews case studies in using instrumentation to support improved global development program deployments.

3.1 Introduction

The first sailors navigated by dead reckoning—a linear extrapolation of their ship's course based on the best-known current and previous locations. Dead reckoning suffered from cumulative error—any mistakes they made accumulated and could lead them dangerously off course.

Celestial navigation vastly improved sailing safety and reliability, by providing reliable and repeatable location determination. Ship captains were no longer blindly reliant on past decisions and used this feedback to make course corrections. However, after a few days of overcast skies there was a tendency for ships to drift.

If we move ahead a few centuries, the advent of GPS chips now provides almost universal and instantaneous location determination for ships, airplanes, and Ubers, based on rapidly triangulating the time delays of speed-of-light transmissions from orbiting satellites.

Such state-of-the-art navigation and control systems are not infallible, however. In 2017, the US Navy Destroyer John McCain collided with an oil tanker, killing sailors. Blame for this accident was placed both on human shoulders and on the poorly designed feedback system, which confused sailors as to the direction and speed of the ships in the minutes before the disaster.

Feedback-supported course correction is used in most domains, from biological sensory feedback to tracking website clicks. We rely on feedback because we generally, without embarrassment or pride, accept that our first judgments are often flawed and require regular updates.

Most consumer and business-to-business-oriented engineering companies rely on customer and competitor feedback to innovate, iterate, and tune their product or service performance and pricing.

However, in global engineering and development the consumer of the product or service often is not the customer—they are the "beneficiaries." Instead, the customer is an international donor, such as the World Bank, USAID, UK DFID, or UNICEF, and some interface with

E. Thomas, *The Global Engineers*, Sustainable Development Goals Series,
https://doi.org/10.1007/978-3-030-50263-8_3

national government ministries. Program design and planning often occurs months or years in advance of deployment, and improvements or corrections identified while the work is actually occurring often cannot be accommodated by contract "change requests" and "deviations."

In the absence of more intuitive customer feedback-driven business models, global engineering and development initiatives often substitute "Theories of Change" and "Logical Frameworks" to describe their program objectives, activities, desired results, and impact and measurement approaches.

These measurement approaches fall broadly into two categories—program monitoring and evaluation (M&E), and impact evaluation (IE). M&E approaches are often modeled closely on the program Theory of Change and Logical Framework, while the measures themselves have more to do with timelines, milestones, and counting of items or activities delivered. Impact evaluation, in contrast, usually largely ignores program design and delivery, and attempts to directly measure the intended impact of the program. The gold standard in impact evaluation is randomized controlled trials, modeled after medical drug trials.

The M&E and IE approaches have demonstrated weaknesses, however. M&E, while often occurring nearly concurrently with program delivery, does not often measure parameters directly relevant to the program motivations. Impact evaluations, while more rigorous, require closely defined programs without allowance for course correction; they are also costly, and often take years to collect and publish their results.

Neither measurement approach is well suited to iterative program evolution or accountability. Furthermore, program funders, implementers, and evaluators—the professionals of the global development sector—are not generally accountable for the outcomes, which are often reported either years after program implementation, when almost everyone (besides the beneficiaries) has moved on.

Measurement and feedback are increasingly being incorporated into global engineering and development financing and programming. Organizations including Feedback Labs (www.feedbacklabs.org), Measure What Matters (www.measurewhatmatters.info), and Proof of Impact (www.proofofimpact.com) are emphasizing better measurement, feedback, program improvements, and, critically, working to support the linking of program funding to measured outcomes.

Within this ecosystem, there is room to incorporate near-time measurement of program, product, and service performance, and feed these measures back to responsible parties (i.e., donors, implementers, governments, communities) in a way that enables iterative testing and improvement.

3.1.1 Sensors Supporting Services

A recent advertising campaign suggested that "Good mornings are better when your alarm clock talks to your coffee maker." Your day improves from there, as your refrigerator reminds you to buy beer, your thermostat gets the house cozy before you get home, and your car guides you away from traffic jams. Such services are indicative of the advent of the Internet of Things (IoT), a market anticipated to reach around 25 billion connected devices in 2020. However, IoT is not limited to making marginal quality of life improvements for suburban America. It can also help bring critical lifesaving services to rural Africa, Asia, and other emerging economies.

While some people worry about warm coffee and cold beer, billions of others in developing countries are preoccupied with finding clean drinking water, safe sanitation, and reliable energy from sources other than firewood and kerosene.

For instance, while the proportion of people with access to water and sanitation has crept up only slowly since 2000, access to cellular data networks has exploded. In Africa alone, the GSM Association estimates that 125 million people who lack access to safe drinking water have mobile coverage. Globally over a billion people within the reach of cellular networks have unsafe sanitation.

The addition of electronic sensors, connected over cellular networks to Internet databases available globally, can help draw attention to, and incentivize, fixing what we might call the Internet of Broken Things.

These sensor technologies can be used both operationally, to monitor any given product or project, or as part of a research sample. Data can be logged locally for manual retrieval or transmitted over short ranges to nearby enumerators, or remote operators and researchers via Wi-Fi, cellular, and satellite networks. Given the remote and power-constrained environments involved, and the high degree of variability in the characteristics of fixed infrastructure, including age, materials, quality, servicing, and functionality, any electronic sensor-based solution would have consisted either of bespoke engineering, or compensate for these complexities through analytics. For example, a conventional flowmeter designed for a rural borehole water distribution scheme would have to address pipe diameter, material, pressure, depth, thread type, and other characteristics that require custom engineering and plumbing. Instead, a nonintrusive ultrasonic flowmeter may be more easily adapted for a variety of water schemes.

Our research group in the Mortenson Center in Global Engineering at the University of Colorado Boulder has iteratively developed and applied electronic sensors within global development settings to support both health studies and the provisioning of basic services. Some of these technologies have been commercialized through a social enterprise, SweetSense Inc. We are also embracing high-tech aerospace tools. Swarm, a company founded by Sara Spangelo and Ben Longmier, is launching small satellites to provide affordable, global data access for IoT devices. Our team is working with Swarm to provide global, cost-effective environmental monitoring of critical air and water resources.

While these electronic sensors are hardly comprehensive or sufficient measurement tools, they provided some novel insights that can help measure and refine interventions.

We have designed sensors to improve service levels and operations and maintenance practices for water infrastructure (Nagel et al. 2015; Turman-Bryant et al. 2019), to inform stakeholders about the use of sanitation technologies (O'Reilly et al. 2015; Sinha et al. 2016; Delea et al. 2017), to understand actual user behaviors with new household products (Thomas et al. 2013, 2016), to support measurement of the health impacts of interventions (Kirby et al. 2019), to establish new methods for higher-efficiency sanitation service delivery, and to examine water collection behaviors in drought-prone areas (Thomas et al. 2019).

Along the way, we have tried to strike a delicate balance between adopting popular technologies—winning grants and contracts to support our approach—and demonstrating that these technologies are only useful if the data they generate is actionable and acted upon, exclusively within complex institutional and community systems. We have simultaneously developed and promoted contributions based on the "Internet of Things," "machine learning," and "remote sensing," while working to improve the provisioning and sustaining of basic water, sanitation, transportation, and energy services.

We have also had to navigate the global development, university research, and technology start-up communities. Each has its own ethos, formulas, and expectations that are occasionally aligned, but are more often in conflict.

We have been supported by grants and contracts from NASA, the World Bank, the National Science Foundation, USAID, the Moore Foundation, the Gates Foundation, the GSM Association, and others. However, we have failed to attract venture capital—a decision by investors that is entirely rational given our complex, slow growth, and mission-oriented market.

Despite the encouragement of communities focused on "information and communication technology for development," social entrepreneurship, impact investing, and public–private-partnerships, the real viability of these initiatives is more sobering. Both global development and technology development are expensive, and neither can progress rapidly in economically suppressed geographies. Worldwide, we have gotten used to cheap consumer electronics. Even sophisticated smartphones are accessible to billions of people. This progress has led to a perception-reality gap in terms of what people think electronics should cost.

With SweetSense, we have focused almost exclusively on applications where sensors may add value to poverty reduction interventions in developing countries. Designing, prototyping, refining, commercializing, and servicing satellite and cellular-connected sensors deployed in remote and challenging environments are expensive, especially when product volumes are limited by the target market. The engineering and staff time must be spread over just a few hundred or thousand products, instead of millions of smartphones.

3.1.2 Designing for Improved Service Delivery

Typically, product design targeted for application in low-income settings has adopted a consumer-oriented model. These approaches have focused on individual- and household-level beneficiaries, with an emphasis on participatory, human-centered co-design methods. Sometimes referred to as "design for the bottom of the pyramid," "design for the developing world" (Lucena et al. 2010), or "social sector human-centered design" (Mitcham and Munoz 2010), these approaches often frame the end-beneficiary as a customer, even when they are not, in fact, retail consumers. The intent is to treat beneficiaries as engaged co-designers in order to develop products that "support development of resource-poor individuals or enhance their capabilities" (Jagtap 2019). These approaches attempt to redress the reality that the majority of the world's engineering effort flows toward product design for the most affluent fraction of the global population.

Our sensor design approach focuses instead on supporting providers of improved water, sanitation, and energy services. This approach prioritizes partnership with governments, utilities, and implementing agencies as relevant stakeholders in this design process. Our design process includes working with service delivery partners, field-based co-design partners, and embracing rapid prototyping, field testing, and product development (Fig. 3.1).

As described in (Sharpe et al. 2019), this approach embraces some components of the emerging consensus around beneficiary co-design, while also maintaining a more traditional relationship between the target product user and the designer. We use iterative design, rapid prototyping, and small-batch manufacturing techniques popularized by the aerospace, automotive, and biomedical industries; we also employ modular product design approaches (Wasley et al. 2016), remote design approaches enabled by advanced computer-aided design (CAD) and communication technologies, and customer feedback systems similar to the human-centered design approach (IDEO 2014).

3.1.3 Water and Sanitation

Efforts to measure the adoption of and compliance with key water and sanitation technologies—such as latrines, water pumps, and water filters—have often relied on surveys and field observations.

Despite what some consider to be the gold standard in WASH monitoring and evaluation, structured observation still has substantial limitations. For one, implementation is resource intensive, requiring time, labor, and funds. Quality enumerators and comprehensive training are required to limit bias due to respondent reactivity and observer differences. This is especially important when the measurement is fairly subjective—inclusive of not only self-reported behaviors and proxy indicators, but also those dependent on observer interpretation—as this has been shown to bias effect estimates (Wood et al. 2008).

Similar to survey reactivity, the act of repeated observation has been shown to augment the behavior of interest, a process known as the Hawthorne effect (Zwane et al. 2011).

While the new SDG indicators for WASH build on established indicators, thereby providing some continuity that will help ongoing

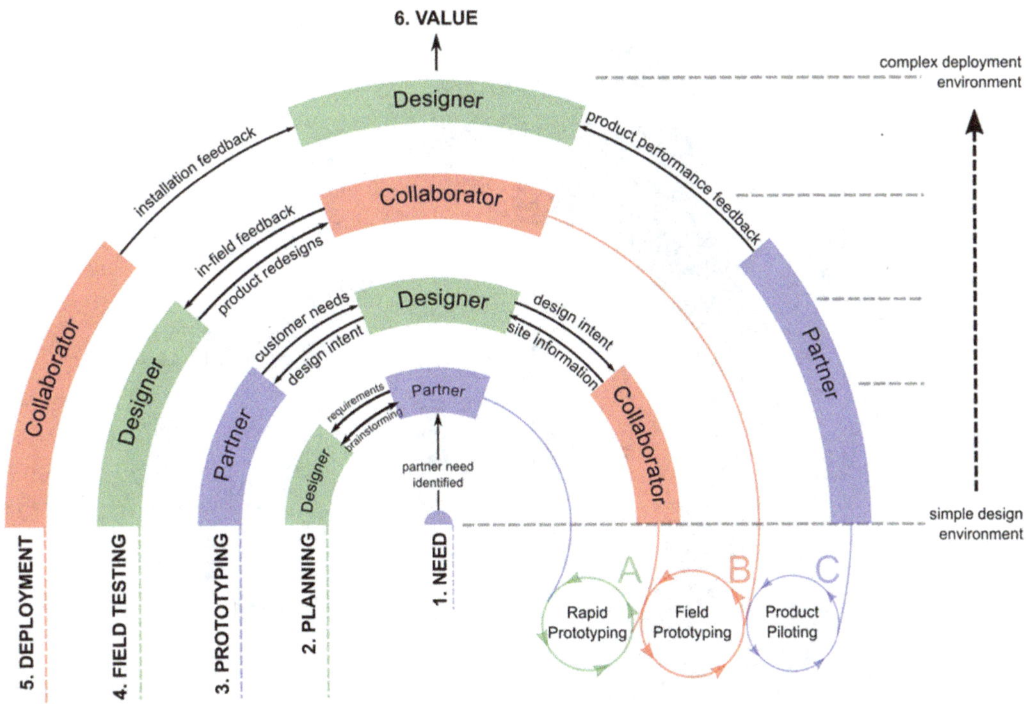

Fig. 3.1 Product design for service delivery approach, including the roles of each design stakeholder and the feedback between them (Design Phases 1–5). Iterative product design processes occur between product development phases (Iteration Phases A–C) (from (Sharpe et al. 2019))

monitoring efforts, they also move beyond them. This places considerable demand on national household surveys to develop new data collection methodologies (Fig. 3.2).

3.1.4 Water Quality Surveillance

Since their promotion in World Health Organization (WHO) monitoring guidelines in 1993, sanitary inspections have become a common component of global water quality surveillance programs. They were developed to provide a rudimentary comparable method for quantifying risk factors that can contribute to microbiological contamination of water sources.

In West Bengal, India, we worked with the Public Health and Engineering Department to conduct sanitary inspections and test for thermotolerant coliforms (TTCs), a fecal indicator bacteria, in 7,317 unique water sources.

Sensors measured the temperature, time, duration, and location of these water quality tests, allowing us to not only accurately correlate sanitary survey scores to water quality results, but also to perform quality control measures during data collection, compare the results to rainfall data, and analyze the data for geospatial and temporal variabilities. We then compared the water quality results to the sanitary inspection scores for each water point.

Our results indicated that the ability of the sanitary inspection score to identify TTC-contaminated sources is poor (Snoad et al. 2017). Across all source types, the sensitivity of a high/very high sanitary inspection score for TTC contamination (TTC > 1 CFU/100 mL) was 29.4% and the specificity was 77.9%, resulting in substantial misclassification of sites when using the established risk categories.

These findings suggest that sanitary surveys are inappropriate screening tools for identifying

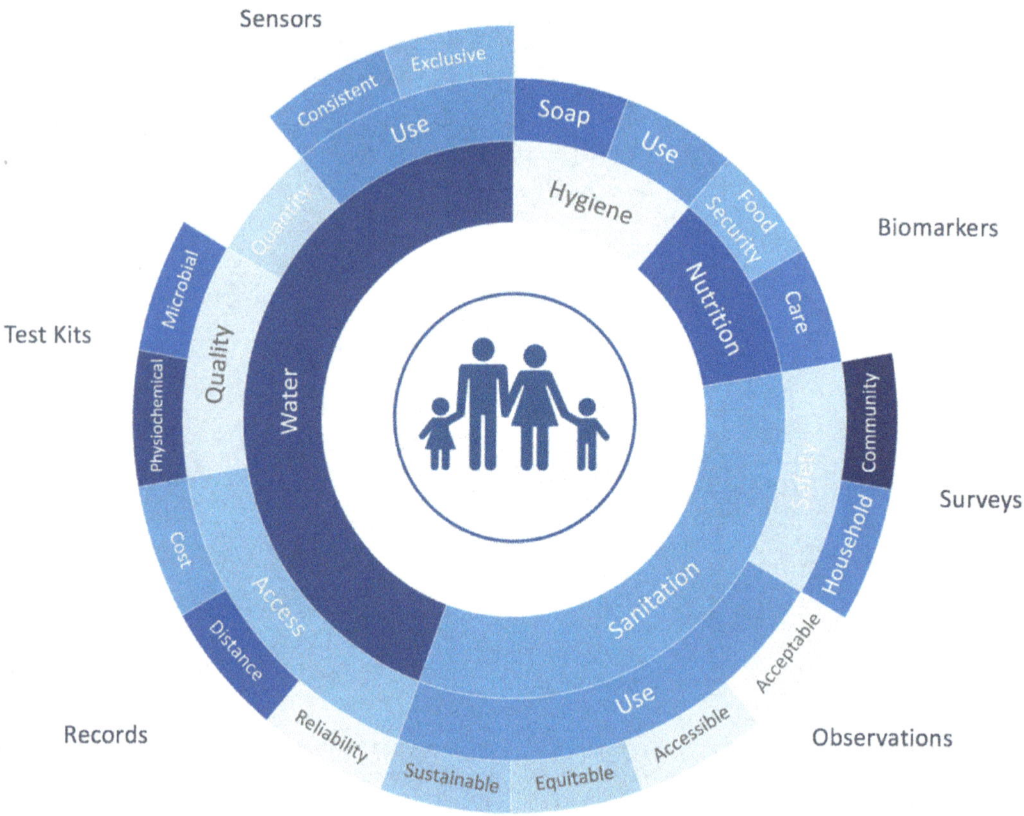

Fig. 3.2 The WHO and UNICEF Joint Monitoring Programme provides indicators for measuring progress toward United Nations Sustainable Development Goal 6, "Clean water and sanitation." These indicators, such as consistent and exclusive use of clean water, have a variety of potential measurement tools including biomarkers, sensors, surveys, observations, and institutional records

TTC contamination at water points and that water quality risk assessment feedback may need to focus on direct water quality testing.

3.1.5 Sanitation Behaviors

Several health and behavioral studies have employed the Passive Latrine Use Monitor (PLUM), first developed at the University of California at Berkeley and commercialized by SweetSense Inc. The sensor uses a simple passive infrared motion detector to identify warm-bodied movement in a latrine stall. Structured observation-validated algorithms then process these movement data into discrete use estimates. These estimates have been shown to be more reliable in individual households than in community latrines where lines may form. In some cases, the sensors include radio-frequency identification (RFID) readers to enable latrine-servicing requests, as demonstrated by Sanergy Inc. in Nairobi, Kenya (Turman-Bryant et al. 2018).

In one study in Bangladesh, these sensors demonstrated that survey data can result in significant over-reporting of latrine utilization. Across 1,207 households randomly selected through 52 village committees in Bangladesh, the mean four-day self-reported latrine use was 32.8 uses, while sensor analysis suggested 21.7 uses on average, indicating a more than 50% exaggeration of latrine use. At the low end of the regression model, the intercept suggests that many households report using latrines when in fact no use is detected (Delea et al. 2017). Similarly, in a recent cross-sectional study among 292 households in 25 villages in India, these sensors were installed for two weeks and compared with household surveys. The mean reported daily use was nearly twice that of the sensor-recorded use, with moderate agreement between daily reported use over the previous 48 h (Sinha et al. 2016).

These findings suggest that the conventional methods of evaluating the use, and subsequent attributable impact, of sanitation interventions may not accurately reflect real behaviors.

3.1.6 Household Water Treatment Behavior

Subject reactivity, when research participants change their behavior in response to being observed, has been documented highlighting the effect of human observers. In Rwanda, we also evaluated whether awareness of these sensors would impact household use of water filters or cookstoves.

We conducted a cluster randomized controlled trial in Rwanda among 170 households (70 blinded to the presence of the sensor, 100 open) testing whether awareness of an electronic monitor would result in a difference in weekly use of household water filters and improved cookstoves over a four-week surveillance period.

We observed a 63% increase in the number of uses of the water filter per week in the group that knew the sensor was installed, compared to the group that did not. This difference declined to 55% a month later. We also observed that the volume of water used differed significantly between the groups, and when compared to survey-based estimates. Figure 3.3 illustrates these findings (Thomas et al. 2016).

There were no significant differences in the number of stove uses per week between the two groups. For both filters and stoves, use decreased in both groups over four-week installation periods.

This study suggests that sensors might function as a means to reinforce healthy behaviors. Yet, it raises a host of issues for researchers. For instance, to accurately measure behavior, sensors may need to be hidden. And that could create privacy and ethical questions.

3.1.7 Household Air Pollution and Clean Cook Stoves

Building on our findings which suggested sensors could be used to directly influence household product adoption behavior, we next applied a human-centered design approach to the development of an air quality monitoring and feedback device. The devices were developed with the assistance of households in Rwanda, and then deployed as part of the NIH and Gates Foundation-sponsored Household Air Pollution Intervention Network (HAPIN) trial. HAPIN, a health efficacy trial, provided LPG cookstoves to households and facilitated their exclusive use to determine the maximum possible health benefits that can be achieved by eliminating biomass cooking. Our technology was designed to help remind households to exclusively use their LPG cookstove.

Through a consultation process we found that households may be reminded to exclusively use their LPG stove through messaging that highlights the health and environmental impacts of cooking practices. We also found that individuals may react to both auditory and visual feedback.

Based on these insights, we designed and validated a system linking air particulate monitoring with persistent visual feedback and a dynamic audio alarm. The design is shown in Fig. 3.4.

3.1.8 Fixing Hand Pumps

While billions of dollars are spent annually on water and sanitation interventions in developing countries, the actual results are often short-lived. Oxford University estimates that 30–50% of the

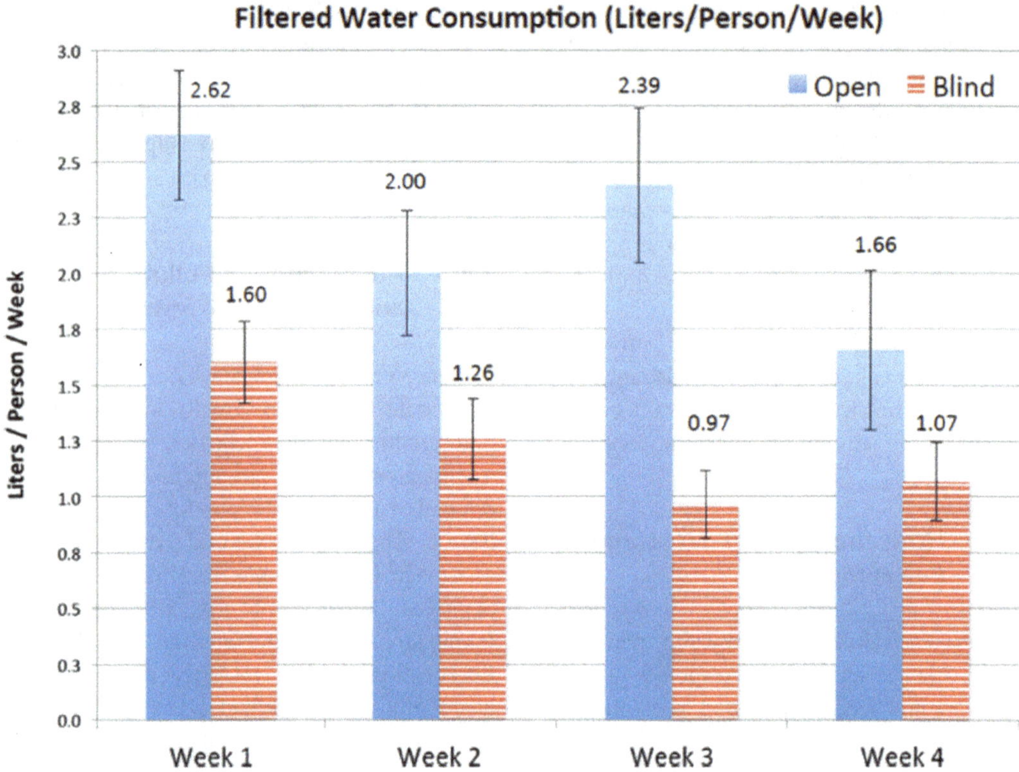

Fig. 3.3 Estimated daily volume (liters) of filtered water consumed per person, per week between households that were aware of a sensor installed in their filter (Open) and households that did not (Blind), over four weeks (reprinted with permission from (Thomas et al. 2016). Copyright 2016 American Chemical Society

millions of water pumps installed in Africa are broken within the first 18 months (Foster 2013). The fundamental reason for these seemingly counterproductive statistics is that the emphasis is overwhelmingly on new installations rather than sustained water services.

In 2014, we worked on a project to install about 180 sensors in rural water pumps in Rwanda. The purpose was to identify pumps that were broken in order to dispatch repair teams. According to a survey, before the sensors were installed some 44% of the area's pumps were broken at any given time, and it took an average of about seven months to get a pump repaired. In three experimental groups, the nominal maintenance model was compared against a "best practice" circuit rider model and an "ambulance" service model. In only the ambulance model was the sensor data available to the implementer, and used to dispatch technicians.

Chantal Iribagiza, a Rwandese engineer and Mortenson Center Ph.D. student, helped lead this study. Chantal's career path is shared in Chap. 5 of this book.

Over the study period, we found that the nominal maintenance group had a median time to successful repair of approximately 152 days, with a mean per-pump functionality of about 68%. In the circuit rider group, the median time to successful repair was nearly 57 days, with a per-pump functionality mean of nearly 73%. In the ambulance service group, the successful repair interval was nearly 21 days with a functionality mean of nearly 91% (Nagel et al. 2015) (Fig. 3.5).

Fig. 3.4 Air quality monitoring and feedback system. The text says "Clean Stove, Good Health" in Kinyarwanda. The grey box in the upper right measures particular matter (PM2.5) and the lungs reflect the current air quality. When the particular matter in the air exceeds a threshold set during a biomass cooking event, a melody plays to remind households to exclusive use their LPG cookstove. Credit Taylor Sharpe, Chantal Iribagiza, Daniel Wilson

An indicative cost analysis suggests that the cost per functional pump per year was roughly similar between the three models. However, the sensor producing and servicing costs were considerably higher during this study than expected at scale.

In Rwanda, the sensor data provided an opportunity to rapidly react to a broken handpump. We then deployed improved sensors on handpumps in Kenya, and leveraged statistical machine learning to try and predict when a handpump would fail, with the objectives of proactively engaging in preventative maintenance and eliminating downtime.

In Kenya, we used sensor data from 42 handpumps observed over 14 months. The baseline mean functionality was 70%. We demonstrated that this could be improved to greater than 99% using our machine-learning model, with just two dispatches per pump-year paired with a one-day delay between notification and dispatch (Wilson et al. 2017).

Our sensors collected the time and magnitude of vibrations and pumped water pressure. To these data, we added features including information on the date, day of week, and location of the pump. We collected actual failures recorded by our sensors and confirmed by the implementers, in order to calibrate and cross-validate our predictive algorithm.

Figure 3.6 illustrates our modeled true positive rate, true negative rate (TNR), positive predictive value (PPV), and negative predictive value (NPV) plotted as a function of a probability threshold that can be selected by an implementer.

3.1.9 Reducing Drought Emergencies in the Horn of Africa

Climate change is a regressive influence and a rapidly emerging challenge for global engineers. Climate change is now impacting rainfall patterns, agricultural yields, and even disease incidence. Malaria, which has killed more people in human history than any other cause, had one saving grace—the parasites were unable to survive in mosquitoes above 5,000 feet because it was too cold. Population centers like Nairobi in Kenya arose because people discovered that they did not become sick there. In the past few years, persistent malaria has taken root in Nairobi as the planet warms.

Changing rainfall patterns are making wet places wetter and dry places drier. In those dry places, reduced and changing rainfall can cause crop failures. Globally, a quarter of the world's population is facing water stress (Fig. 3.7).

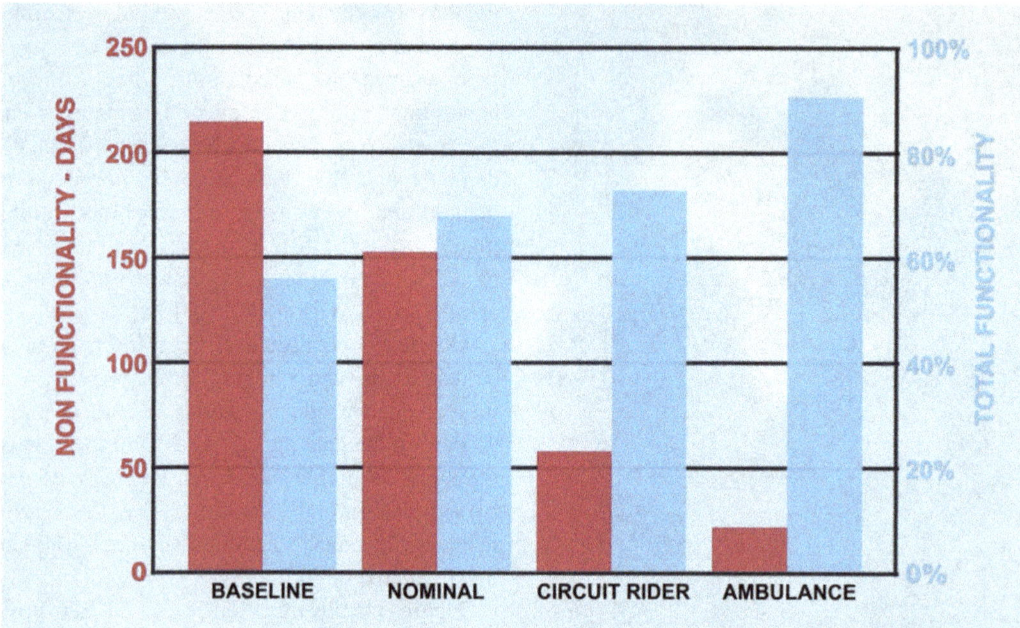

Fig. 3.5 Number of nonfunctional days between repairs (red) and average percentage of functional days for handpumps in rural Rwanda in a baseline survey, and as measured in three maintenance models: a nominal group wherein repairs were scheduled when feasible, a circuit rider model where dedicated technicians routinely visited pumps, and an ambulance model where sensor data triggered repairs. Sensors monitored pump functionality and use in the nominal, circuit rider, and ambulance groups, but the data were provided to the service provider only in the ambulance group (reprinted with permission from (Nagel et al. 2015). Copyright 2015 American Chemical Society)

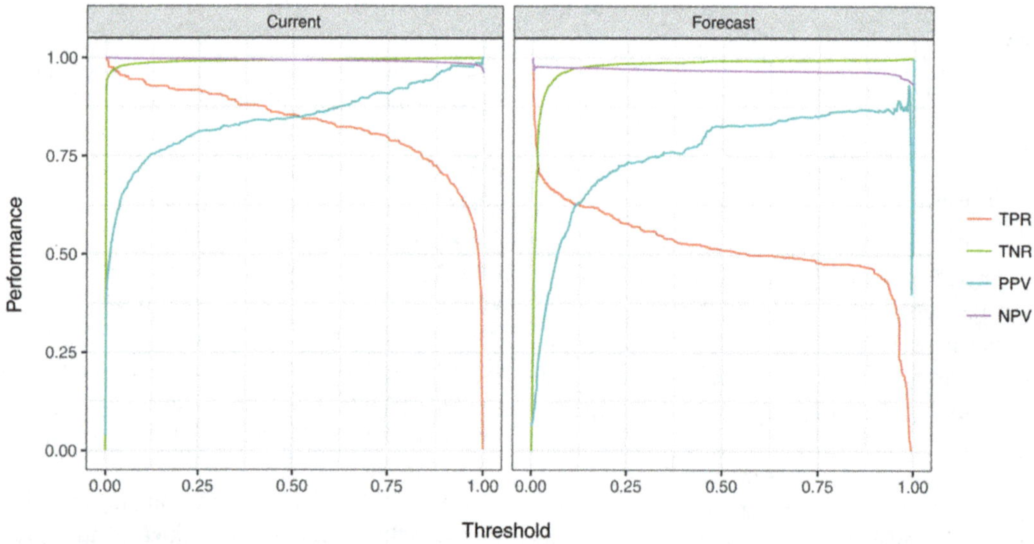

Fig. 3.6 The true positive rate (TPR), true negative rate (TNR), positive predictive value (PPV), and negative predictive value (NPV) are plotted as a function of the learner's probability threshold. This bottom panel illustrates the relationship between learner performance and the implementer-defined probability threshold to decide if a pump is broken (current) or will break (forecast) on any given day (from: (Wilson et al. 2017))

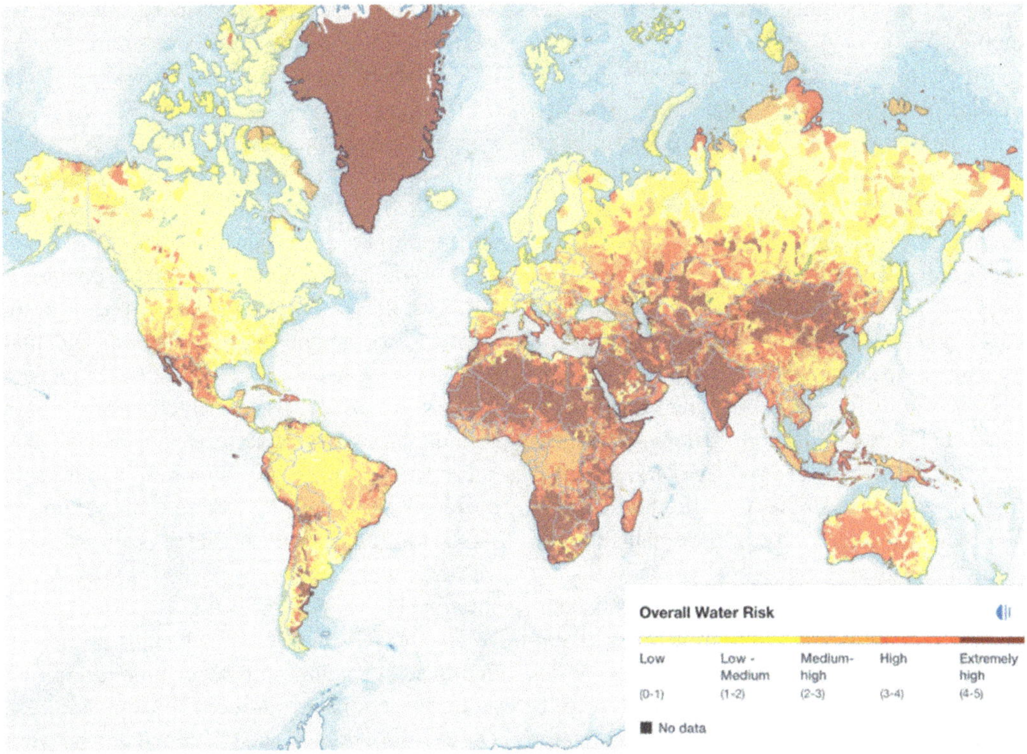

Fig. 3.7 Global water risk. *Source* WRI Aqueduct, accessed on March 25, 2020. https://www.wri.org/aqueduct

Millions of people living in the drought-prone Horn of Africa are facing persistent threats from the lack of safe, reliable, and affordable water year-round (Viste et al. 2013; Shiferaw et al. 2014). The arid regions of Ethiopia, Kenya, and Somalia have experienced growing frequency and severity of drought conditions, which are expected to increase further in coming years (Masih et al. 2014; Ahmadalipour and Moradkhani, 2018). The 2011 drought in East Africa caused food shortages for over 10 million people across the region and over 260,000 deaths in Somalia alone (Shabelle 2011; Nicholson 2014). The recent 2016–17 drought in Kenya resulted in over 3 million people facing food insecurity (Uhe et al. 2017). Preventable death and malnutrition hits hardest in pastoral communities—UNICEF estimates that there are 19.5 million pastoral people in the Horn of Africa, of whom 40% survive on less than $1 a day (UNICEF 2006).

Drought emergencies occur when reduced rainfall, exacerbated in recent years by climate change (Funk et al. 2015), conspires with limited community capacity and institutional failures to cause dramatic reductions in access to water for people, livestock, and agriculture. This lack of water results in catastrophic crop failures, public health stress, economic shocks, and displacement of people. Historically, responses to drought have been *reactive*, involving international emergency assistance to save lives and livelihoods; these then disappear once the immediate crisis dissipates. This reactive, after-the-fact emergency assistance occurs despite the reality that drought in the Horn of Africa is both cyclical and increasing (Masih et al. 2014; Ahmadalipour and Moradkhani 2018). This increase in frequency means that there is no longer sufficient time for pastoralists to recover their herds and crops in between droughts, thus the destabilizing impact increases with each successive event, leading to vulnerability and insecurity in this complex region of Africa.

In Ethiopia, nearly 8 million people are systematically affected by drought, food insecurity, floods, and conflict. Political strife, exacerbated by the economic and social pressures of drought and food insecurity, has led to armed conflicts in the Afar, Oromia, and Somali regions. In the northern districts of Kenya, where we presently operate (Isiolo, Garissa, Marsabit, Turkana, and Wajir), hundreds of thousands of refugees live in UN camps, fleeing internal instability, including refugees from Ethiopia, Somalia, and South Sudan. In Marsabit, 10,000 Ethiopian refugees live outside of UN camps; in Turkana, over 185,000 South Sudanese live in the Kakuma camp; and in Garissa, 245,000 Somali refugees live in the Dadaab camp.

In parts of this region, these economic, political, and refugee stresses are further exacerbated by al-Shabab, which openly operates water trucking cartels, blocks food aid, and recruits from communities facing the economic hardships of drought.

Drought-driven humanitarian emergencies can be prevented if groundwater is made reliably available at strategic locations during cycles of water stress. Over the past decade, countries and international donors have spent millions of dollars to install thousands of groundwater pumping stations, known as *boreholes*, in the region in the hopes of reducing drought emergencies. Unfortunately, these investments are failing to improve the dire situation in the region largely because these boreholes break down during the periods of greatest need (Hope et al. 2012; Nagel et al. 2015; Foster et al. 2018).

Doris Kaberia, a Kenyan expert in food security and pastoral livelihoods, and Petros Birhane, an Ethiopian agricultural engineer and disaster relief expert, have dedicated their careers to building resilience in these regions of Africa. Petros and Doris' share their stores in Chaps. 8 and 9 of *The Global Engineers*.

A new approach to the use of strategic boreholes is needed to end drought emergencies in the Horn of Africa. With Doris, Petros, and other partners, we are working with the governments of Ethiopia, Kenya, and Somalia, and the Millennium Water Alliance—a consortium of many of the largest water-engaged international nonprofit organizations—to develop an integrated, systematic approach called the Drought Resilience Impact Platform (DRIP).

DRIP will empower institutions and communities to take coordinated action to maintain water availability during drought conditions. DRIP applies a comprehensive systems design approach to integrate early detection and planning tools with proactive groundwater management, in order to enable drought-prone communities to become proactive and effective managers working to prevent these humanitarian crises. We will replace reactive and expensive short-term assistance measures such as water trucking with a framework for proactive and sustainable drought resilience.

The DRIP theory of change posits that improved drought indicator monitoring will produce actionable drought impact forecasts. These will inform and strengthen networks including communities, governments, and international organizations, which can then collectively ensure water availability and end drought emergencies (Fig. 3.8).

The DRIP theory of change includes performance-based financial incentives to support the prioritization of cost-effective water pump management. We are currently demonstrating this important pillar in Kenya in partnership with Proof of Impact. Our sensors are reporting functional pump-days to buyers of "impact events" who are now providing funding to local water utilities on a monitored performance basis (https://proofofimpact.com/events/provide-stable-water-access).

The USAID supported Sustainable WASH Systems Learning Partnership (SWS) led by the University of Colorado Boulder, has recently identified key factors in enabling effective water service delivery. These include monitoring capacity, technical capacity, policy and regulatory oversight, and government-led financial incentive programs. Similarly, several factors have been identified that influence water user payments. Most critically, users are more likely

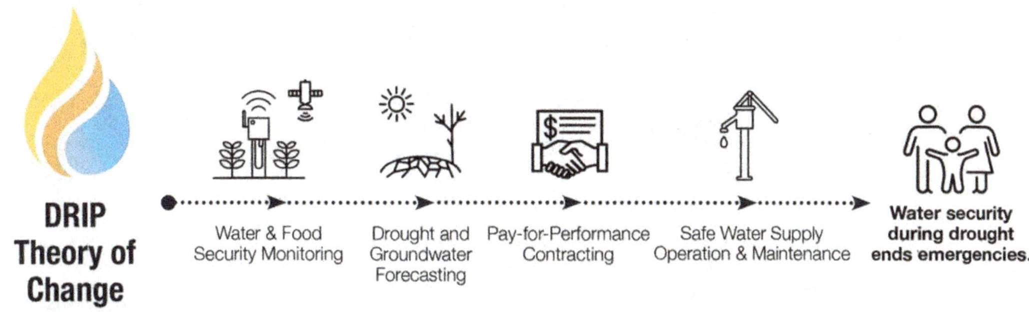

Fig. 3.8 The drought resilience Impact platform theory of Change which posits that improved food and water security monitoring can produce more actionable drought, water demand and food security forecasts. These forecasts can in turn be used to pinpoint water security actions, including water pump repairs, and ensure water security through performance-based funding and contracting

to pay for water services if reliable and fast maintenance and repairs occur (Sustainable WASH Systems Learning Partnership 2020). These water service and water user payment factors are relevant to the DRIP theory of change and deployments.

Dan Hollander, an American civil engineer and former USAID foreign service officer, leads the SWS program. Dan's experience working in low, middle, and high-income countries as an engineer is shared in Chap. 10.

This work builds in part on our ability to monitor groundwater pump functionality and use, and link these data with institutions who can take action to maintain water access. Using our satellite and cellular-connected sensors, we currently monitor all of the government-designated drought response groundwater pumps in northern Kenya and nearly all motorized boreholes in Afar, Ethiopia, with expansion foreseen into Somali, Ethiopia. As of early 2020, we are continuously monitoring the water supplies of over 3 million people on a daily basis (Fig. 3.9).

We have demonstrated the strong interconnectedness between rainfall and groundwater use in this region. In a recent study, we examined three remote-sensing data sets against in situ sensor-collected groundwater extraction data from 221 water points, serving over 1.34 million people across northern Kenya and Afar, Ethiopia, between January 1, 2017 and August 31, 2018. In models containing rainfall as a binary variable, we observed an overall 23% increase in borehole runtime following weeks with no rainfall, compared to weeks preceded by some rainfall. Further, a 1 mm increase in rainfall was associated with a 1% decrease in borehole use the following week (Thomas et al. 2019). When surface water availability is reduced during the dry seasons, groundwater demand increases. Our findings emphasize the urgent need to maintain functionality of groundwater boreholes in these regions, which often suffer drought-related emergencies (Figs. 3.10 and 3.11).

Our data are being used by local communities, regional governments, and national and international donors to reduce repair intervals. This means that communities have more reliable access to water during dry seasons.

In Kenya, the National Drought Management Authority reviews our dashboard weekly and, using the data, works with county governments in northern Kenya to prioritize pump repairs. In Ethiopia, our partners—including DT Global, IRC WASH, and mWater—are working with the Afar and Somali regional water bureaus to adopt the use of these data to better manage water systems.

Working with the NASA and USAID SERVIR program (www.servirglobal.net), geospatial experts in Kenya, and the Famine Early Warning Systems Network (FEWS NET), we also use satellite-based rainfall estimates to predict if a nonused pump is either broken, or simply not being

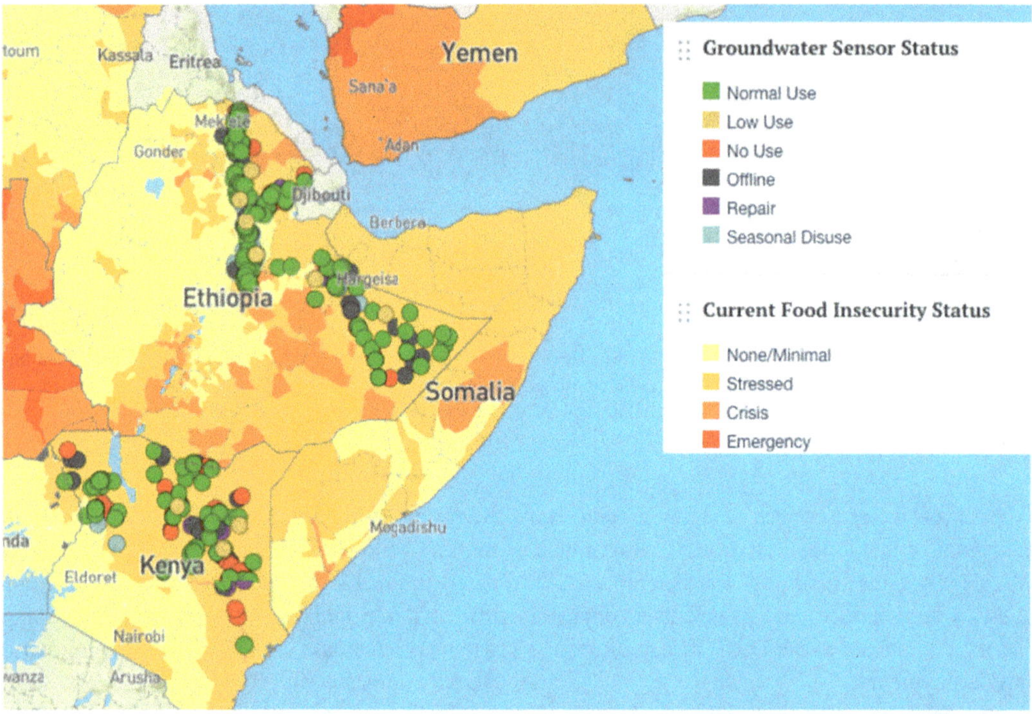

Fig. 3.9 Satellite and cellular-connected groundwater pump monitoring network and the Famine early warning systems network food insecurity status for December, 2019. *Source* prepdata.org, accessed on March 25, 2020

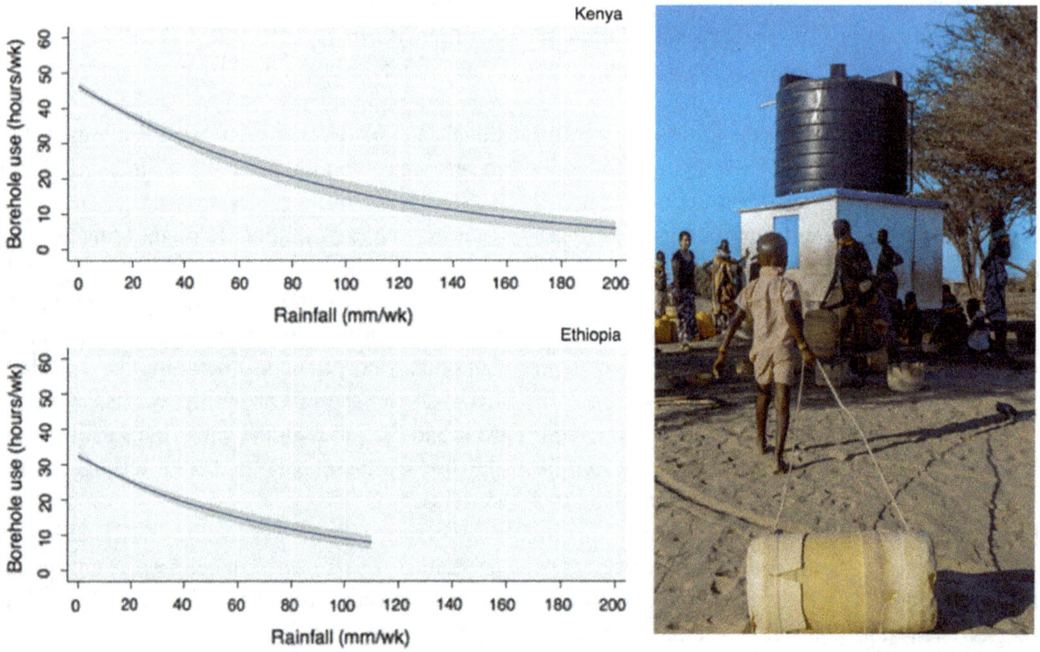

Fig. 3.10 Left—Modeled borehole runtime against weekly rainfall (from: (Thomas et al. 2019) right—An example of a groundwater borehole storage system in Turkana, Kenya

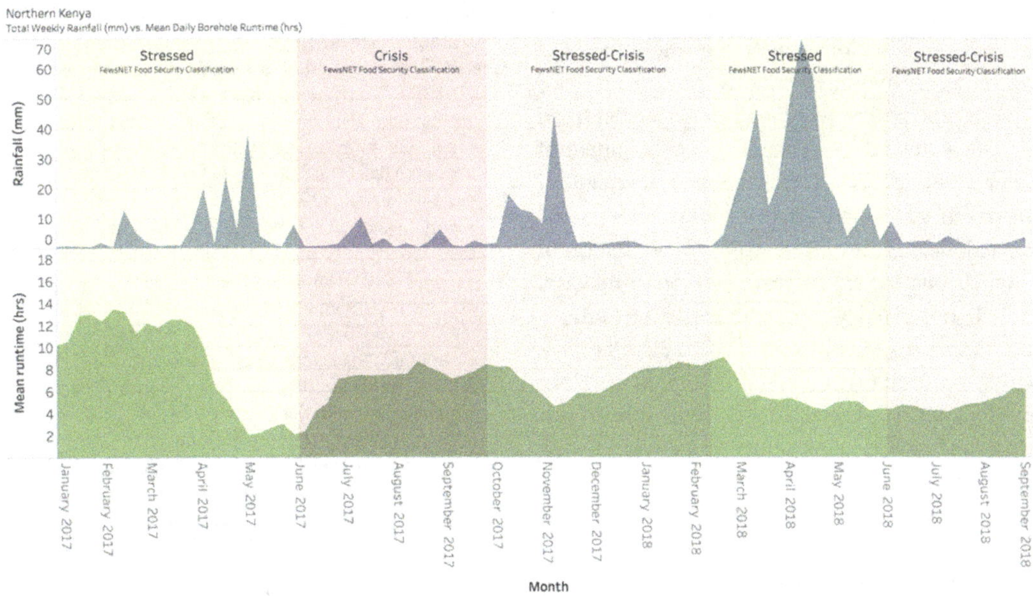

Fig. 3.11 Rainfall (top) as detected with satellite-based remote sensing, and groundwater use (bottom) as recorded with electronic sensors. The colored overlay shows FEWS NET food security classifications for the designated periods in northern Kenya (from: (Thomas et al. 2019))

used because surface water is available, and further link our groundwater use estimates to forecasts for water demand and food security.

As we have seen in Kenya and Rwanda, effectively using real-time data to improve water delivery requires the incorporation of this information into local water management policies and practices. However, little study has been made of the barriers and facilitators to utilizing real-time data for water management in low-income settings. As a consequence, we have limited guidance at our disposal for the development of effective strategies to promote adoption and implementation of sensor-based technologies, in order to improve water service delivery.

A useful framework for determining the impact of sensor-enabled boreholes on water delivery, while simultaneously evaluating and refining methods to implement these data into local water management practices, is the hybrid effectiveness-implementation design (Curran et al. 2012). This class of study design combines rigorous summative evaluation of the effectiveness of a given intervention with formative evaluation of the progress and effectiveness of intervention implementation. Importantly, unlike the traditional randomized clinical trial, this approach allows information from formative evaluation to be utilized during the conduct of the study to optimize implementation and create a framework for assessing the impact of evolving implementation strategies.

Our ongoing studies use an interrupted time series (ITS) design to evaluate whether the provision of local and national institutions with real-time data from the instrumented boreholes, in conjunction with training and capacity activities, results in improved water service delivery.

The ITS design is widely regarded as the most rigorous of the experimental designs in which randomization is infeasible (Anderson-Cook 2005). ITS designs require collecting data on the outcome of interest at multiple time points both before and after an intervention in order to test whether the intervention has an effect significantly greater than the baseline (pre-intervention) secular trend (Bernal et al. 2017). Although most commonly used in the context of single time series, the ITS design can be readily adapted to situations involving multiple sites and/or successive interventions.

A recent policy review in *Science* identified four steps toward the sustainable development of water resources: (1) Improved measurement of watershed status and water use via remote sensing and local electronic sensors, (2) Valuation of water resources, (3) Improved decision-making, and (4) Improved water governance institutions and incentives (Garrick et al. 2017). Our proposed approach to drought mitigation in the Horn of Africa incorporates the tenets described in this paper, and extends their utility to reducing water-related humanitarian crises.

DRIP does not presume to solve the underlying causes of drought, which include climate change, population growth, and increasing agricultural demands. However, we do possess the capability to more effectively manage the available local and national resources to avoid the humanitarian and societal emergencies that often result from drought. Here, we propose leveraging systems understanding with technical and policy expertise to adapt to this changing and dynamic environment. In this way, we can turn drought into a manageable condition that can be sustainably planned for, managed, and mitigated by the communities and countries with the most at stake.

Taken as a whole, these example programs illustrate the value of combining engineering, measurement and actions to improve access to basic services, and require the participation of many levels of institutions from communities to governments to international donors.

References

Ahmadalipour A, Moradkhani H (2018) Multi-dimensional assessment of drought vulnerability in Africa: 1960–2100. Sci Total Environ. https://doi.org/10.1016/j.scitotenv.2018.07.023

Anderson-Cook CM (2005) Experimental and quasi-experimental designs for generalized causal inference. J Am Stat Assoc. https://doi.org/10.1198/jasa.2005.s22

Bernal JL, Cummins S, Gasparrini A (2017) Interrupted time series regression for the evaluation of public health interventions: a tutorial. Int J Epidemiol. https://doi.org/10.1093/ije/dyw098

Curran GM, Bauer M, Mittman B, Pyne JM, Stetler C (2012) Effectiveness-implementation hybrid designs: combining elements of clinical effectiveness and implementation research to enhance public health impact. Med Care 50(3):217

Delea MG et al (2017) Comparison of respondent-reported and sensor-recorded latrine utilization measures in rural Bangladesh: a cross-sectional study. Trans R Soc Trop Med Hyg 111(7):308–315. https://doi.org/10.1093/trstmh/trx058

Foster T (2013) Predictors of sustainability for community-managed handpumps in sub-Saharan Africa: evidence from Liberia, Sierra Leone, and Uganda. Environ Sci Technol 47(21):12037–12046. https://doi.org/10.1021/es402086n

Foster T et al (2018) Risk factors associated with rural water supply failure: a 30-year retrospective study of handpumps on the south coast of Kenya. Sci Total Environ. https://doi.org/10.1016/j.scitotenv.2017.12.302

Funk C et al (2015) Assessing the contributions of east African and west pacific warming to the 2014 boreal spring east African drought. Bull Am Meteor Soc. https://doi.org/10.1175/BAMS-D-15-00106.1

Garrick DE et al (2017) Valuing water for sustainable development. Science 358(6366):1003–1005. https://doi.org/10.1126/science.aao4942

Hope R et al (2012) Harnessing mobile communications innovations for water security. Global Policy 3 (4):433–442. https://doi.org/10.1111/j.1758-5899.2011.00164.x

IDEO (2014) The field guide to human-centered design, Igarss 2014. https://doi.org/10.1007/s13398-014-0173-7.2

Jagtap S (2019) Design and poverty: a review of contexts, roles of poor people, and methods. Res Eng Des. https://doi.org/10.1007/s00163-018-0294-7

Kirby MA et al (2019) Effects of a large-scale distribution of water filters on water quality and diarrhoea: a cluster randomized controlled trial in Western Province, Rwanda. PLOS Med 1–24. https://doi.org/10.1371/journal.pmed.1002812

Lucena J, Schneider J, Leydens JA (2010) Engineering and sustainable community development. Synth Lect Eng Technol Soc. https://doi.org/10.2200/S00247ED1V01Y201001ETS011

Masih I et al (2014) A review of droughts on the African continent: a geospatial and long-term perspective. Hydrol Earth Syst Sci. https://doi.org/10.5194/hess-18-3635-2014

Mitcham C, Munoz D (2010) Humanitarian engineering. Synth Lect Eng Technol Soc. https://doi.org/10.2200/s00248ed1v01y201006ets012

Nagel C et al (2015) Evaluating cellular instrumentation on rural handpumps to improve service delivery-A longitudinal study in rural rwanda. Environ Sci Technol 49(24):14292–14300. https://doi.org/10.1021/acs.est.5b04077

Nicholson SE (2014) A detailed look at the recent drought situation in the greater Horn of Africa. J Arid Environ. https://doi.org/10.1016/j.jaridenv.2013.12.003

O'Reilly K et al (2015) Combining sensor monitoring and ethnography to evaluate household latrine usage in rural India. J Water Sanit Hyg Dev 5(3):426–438. https://doi.org/10.2166/washdev.2015.155

Shabelle L (2011) Drought-related food insecurity: a focus on the Horn of Africa. Drought Emergency

Sharpe T, Muragijimana C, Thomas E (2019) Product design supporting improved water, sanitation, and energy services delivery in low-income settings. Sustainability 11(23):6717. https://doi.org/10.3390/su11236717 (Multidisciplinary Digital Publishing Institute)

Shiferaw B et al (2014) Managing vulnerability to drought and enhancing livelihood resilience in sub-Saharan Africa: technological, institutional and policy options. Weather Clim Extremes. https://doi.org/10.1016/j.wace.2014.04.004

Sinha A et al (2016) Assessing Latrine use in rural India: a cross-sectional study comparing reported use and passive Latrine use monitors. Am J Trop Med Hyg 95 (3):720–727. https://doi.org/10.4269/ajtmh.16-0102

Snoad C et al (2017) The effectiveness of sanitary inspections as a risk assessment tool for thermotolerant coliform bacteria contamination of rural drinking water: a review of data from West Bengal, India. Am J Trop Med Hyg. https://doi.org/10.4269/ajtmh.16-0322

Sustainable WASH Systems Learning Partnership (2020) Maintenance approaches to improve the sustainability of rural water supplies

Thomas EA et al (2013) Use of remotely reporting electronic sensors for assessing use of water filters and cookstoves in Rwanda. Environ Sci Technol. https://doi.org/10.1021/es403412x

Thomas EA et al (2016) Behavioral reactivity associated with electronic monitoring of environmental health interventions—A cluster randomized trial with water filters and cookstoves. Environ Sci Technol 50 (7):3773–3780. https://doi.org/10.1021/acs.est.6b00161

Thomas EA et al (2019) Quantifying increased groundwater demand from prolonged drought in the East African Rift Valley. Sci Total Environ. https://doi.org/10.1016/j.scitotenv.2019.02.206

Turman-Bryant N et al (2018) Measuring progress towards sanitation and hygiene targets: a critical review of monitoring methodologies and technologies. Waterlines. https://doi.org/10.3362/1756-3488.18-00008

Turman-Bryant N et al (2019) Improved drought resilience through continuous water service monitoring and specialized institutions—A longitudinal analysis of water service delivery across motorized Boreholes in Northern Kenya. Sustainability 11(11). https://doi.org/10.3390/su11113046

Uhe P et al (2017) Attributing drivers of the 2016 Kenyan drought. Int J Climatol 38(Suppl. 1):e554–e568. https://doi.org/10.1002/joc.5389

UNICEF (2006) Child alert Horn of Africa. https://www.unicef.org/media/files/Child_Alert_HoA_FINAL.pdf

Viste E, Korecha D, Sorteberg A (2013) Recent drought and precipitation tendencies in Ethiopia. Theoret Appl Climatol. https://doi.org/10.1007/s00704-012-0746-3

Wasley NS et al (2016) Experimenting with concepts from modular product design and multi-objective optimization to benefit people living in poverty. Dev Eng. https://doi.org/10.1016/j.deveng.2016.12.002

Wilson DL, Coyle JR, Thomas EA (2017) Ensemble machine learning and forecasting can achieve 99% uptime for rural handpumps. PLoS ONE 12(11): e0188808. https://doi.org/10.1371/journal.pone.0188808

Wood L et al (2008) Empirical evidence of bias in treatment effect estimates in controlled trials with different interventions and outcomes: meta-epidemiological study. BMJ (Clinical research ed.), 336(7644):601–605. https://doi.org/10.1136/bmj.39465.451748.ad

Zwane AP et al (2011) Being surveyed can change later behavior and related parameter estimates. Proc Natl Acad Sci USA. https://doi.org/10.1073/pnas.1000776108

Part I
Global Engineers and Partners

The balance of this book is dedicated to profiles of engineers and other technical professionals who have dedicated their careers to searching for solutions to global development challenges. These individuals were not selected randomly or scientifically. They are all professionals with whom I have worked, and who share diverse views on the proper roles, responsibilities, and opportunities for engineers in global development.

These professionals include Heather Fleming, a Navajo designer and the founder of Catapult Design, Chantal Iribagiza, a Rwandese engineer specializing in rural water supplies, Jean Ntzinda, a Rwandese environmental planner responsible for facilitating foreign corporate and donor-supported programs, Avery Bang, an American civil engineer and CEO of Bridges to Prosperity, Doris Kaberia, a Kenyan expert in food security and pastoral livelihoods, Petros Birhane, an Ethiopian agricultural engineer and disaster relief expert, and Dan Hollander, an American hydrological engineer and former foreign service officer.

Together, these stories introduce the reader to the diverse opportunities and challenges in Global Engineering.

4 Heather Fleming

Abstract

Heather Fleming is the founder of Catapult Design, a nonprofit design firm that "drives services for socially driven clients." Heather was born on the Navajo Nation and grew up a few miles away from the reservation with her family. A graduate of Stanford University's product design program, Heather has dedicated her career to product and service design to reduce poverty and improve livelihoods globally. Heather's career has reflected the challenges in developing products and services designed to address poverty, as described in Chap. 1.

E. Thomas, *The Global Engineers*, Sustainable Development Goals Series,
https://doi.org/10.1007/978-3-030-50263-8_4

Heather Fleming is the founder of Catapult Design, a nonprofit design firm that "drives services for socially driven clients." Catapult Design has won contracts from the Gates Foundation, USAID, the World Bank, and many others to apply the latest design and engineering trends to product and service development in Guatemala, India, Kenya, Peru, Rwanda, and Tanzania.

Heather was born on the Navajo Nation and grew up a few miles away from the reservation with her family. Rivalling the conditions in many developing countries, 30% of households in the Navajo Nation have no running water, while 70% of those that do receive water contaminated with fecal bacteria. As a child, Heather would help her grandfather load 55-gallon drums into his pickup truck to collect water from nearby wind-powered water pumps, which were often broken. As she recalled, "*The government had money to install them, but there was no budget or division of the government responsible for maintenance—not even the NTUA (Navajo Tribal Utility Authority).*"

"*We used that water for everything. We had these big aluminum bowls for hand washing. You'd wash your hands in dirty water because everyone used the same water all day long. My auntie then used that water to do other things, like pour it on the concrete floors to get the dirt off. Bathing was always a challenge. There were no showers, so it was a quick sponge bath at night.*"

"*Warm water was reserved for cooking. My grandpa has this extremely inefficient cast iron stove issued by the Navajo Housing Authority. Most people use either coal, propane tanks, or they chop down wood to heat their homes. This contributes to deforestation on the Reservation.*"

"*In 2009 an NTUA representative told me it costs tens of thousand dollars to connect a home to electricity on the Navajo Nation. Most people can't afford the $30 per month electricity bill.*"

Heather's father, a Health Service doctor, met her Navajo mother while working on the Navajo Nation. "*My mom did not teach us Navajo deliberately. She's one of nine children. My older aunties and uncles, more or less, live traditional lives. My younger aunties and uncles were all taken to boarding schools put in place to wash Native American kids of their culture. My mom has bad memories of being taken from her family, being separated from her brothers and sisters at school, and not being allowed to speak Navajo. I think that that aspect of her childhood affected her decision to not speak Navajo to my sister and I. She decided that teaching my sister and I Navajo would hinder us.*"

"*When I was in high school one of my clan cousins wanted to teach me and two of my cousins, and my mom said it was okay. I learned a few basic things then. The most important thing to know in Navajo is how to introduce yourself with your clans. It allows people to place you because each clan has characteristics that define them and some clans are related. My grandfather lived with us for a while and we couldn't communicate. He would speak Navajo to me and my mom served as a translator. I played the piano for him a lot, so maybe we spoke through music more than anything else.*"

"*We raised ducks and chickens when I was little in addition to rabbits, hamsters, and fish. At one time we had 10 dogs and nine cats. I spent most of my time outside with my dogs on the 10 acres of land my family owned in rural New Mexico. I used to like to hide in the bushes and spy on my family or our nearest neighbor, who lived about a half a mile away. During the summers, I moved rocks in the Arizona heat. The road to my grandpa's house wasn't paved, it was just a bumpy dirt road. My mom and my Auntie Esther were constantly trying to make it easier to drive to the house so in the summers we put rocks on the road. I spent a lot of summers picking up and moving rocks and making mud cakes.*"

Heather took the bus from Vanderwagen to attend high school in Gallup, New Mexico. Nearly everyone on the bus was Navajo. "*Gallup was a multicultural town. Our high school was majority Native American, then a large Hispanic population, and a minority White population. My closest friends in high school were first-generation Nepali, Filipino, Indian, and also Native American and Mexican American."*

On the bus, the other Navajo students thought Heather was an outsider, and called her a *Bilagáana*. "*It means white person in Navajo. I grew up with my Navajo family, so I did not consider myself a Bilagáana at all. But my sister and I both have lighter skin. We look more like my dad than my mom. Those kids on the bus didn't know me, or that my mom is Navajo. That always bothered me. That awkwardness of feeling like you're part of one community, but that community doesn't see you as part of them. And the other community that does see you as part of them, you don't really identify with. You're stuck in between the two.*"

Heather's cousin Genie was a civil engineer working for the Indian Health Service on the Navajo Nation. In high school, Genie encouraged Heather to consider engineering.

"*So, I got in her white government truck and we drove out past Window Rock towards Kinlichee. Halfway there we turned down a dirt road and drove for what felt like a really, really long time. I didn't know where we were going, we were just talking about college and what I was going do after high school. At some point we reached a big water tank, and she stopped the car and said, "I built that, and I designed it." She pulled out these big maps of the reservation. She had little red dots marking all of the houses that she had come across down that road. "I built this well for these families," and then we went to talk with those families. And they all knew her by name.*"

"*After bathing and washing my hands at my auntie's house in the aluminum bowls, I understand how transformational running water is. To learn that my cousin had built a well seemingly in the middle of nowhere, and that her actions meant that families now had running water, was incredible to me. I remember thinking, "Wow, civil engineers are amazing! They're humanitarians, they solve problems, and the things they make have a big impact on people's lives." I wanted to do that too. At the time I wanted to do exactly what she did, without really knowing what she did. I saw her interacting with people and building things that made it so that they didn't have to use dirty water to wash their hands or take a sink bath with a cold basin of dirty water.*"

"*Near the end of my junior year in high school, my dad and I took an Amtrak to California. He took me to see Stanford, UC Santa Cruz, UC Davis, Berkeley. He was trying hard to get me to think about not staying where I was. I ended up applying to Stanford because of a deal I made with him on that trip. My dad has always been an avid sports fan, so my sister and I played sports consistently throughout our childhood. But I didn't want to play anymore, so I told him I would apply to college if I didn't have to play basketball, volleyball, or run track my senior year of high school.*"

"*I stated in my Stanford essay that I wanted to be either a civil engineer or a computer scientist.* Toy Story *came out when I was in ninth grade and I was obsessed with computer animation. There was a $30 book at Waldenbooks about how* Toy Story *was made, and I begged my parents to get it. What stuck with me the most is the explanation of how the animators imagined each character's movement. You know the green soldier men? You know how their feet are stuck to the base? All the animators had taken shoes and nailed them to a piece of wood. And they used that to figure out how green toy solders move when their legs were stuck to something. I thought that was the coolest thing. In retrospect it's an empathy exercise. It's trying to understand if someone has this condition, or has this circumstance, how does that change the way your world works. Literally walking in somebody else's shoes.*"

"*Stanford is extremely manicured, unlike the Navajo Nation. There's a relatively large Native American population there, but I didn't try to take part in that community. I think high school did that to me, and probably those bus rides. I didn't feel like I fit in the Native community, despite being born on the Navajo Nation and growing up with family there.*"

When Heather started her engineering courses at Stanford, she quickly became disheartened by some of the topics on the curriculum. "*My perception of engineering was based on what I saw my cousin do. I thought, 'oh drilling wells and helping families, that sounds fun.' But when I sat through math classes, I thought, 'This is not how I want to spend my life.*'" After a job fair, Heather switched to product design. "*In my first year of engineering classes, we never once talked about people. We didn't talk about the social issues surrounding engineering projects. It was all math and physics. But the product design curriculum is centered around people and human need in addition to engineering. The emphasis is on how you create things—objects, systems, services—for people that are meaningful, and that last. And that appealed to me more than using an equation to arrive at a single answer.*"

During Heather's senior year at Stanford, she tried to tie her product design and engineering skills to the humanitarian challenges she had observed back home. "*The graduating seniors and the first-year master students are expected to present a product design thesis at the end of the school year. You have to conceive of, design, and then implement a product based on a problem, or need, that you've observed in the world. I picked a problem based on my experience on the Rez—the need for easier and more efficient water transport solutions for the families without running water. At the time I didn't realized that it was actually a global issue affecting hundreds of millions of people. I just knew it was a big problem back home. I shared my chosen problem statement with my professor and his feedback to me was that my thesis work should reflect the career opportunities I want to pursue. He more or less told me, "This is not a problem that represents a huge number of people." In retrospect, many of our professors were designing products exclusively for Western markets. And in many ways, as students that's what we were trained to strive for.*"

"*My professor said, 'I just don't see a market for this idea.' It's ironic in retrospect. Today Stanford's most popular class is Design for Extreme Affordability and IDEO has IDEO.org. But at the time they weren't encouraging students to look beyond traditional Western markets for problems to solve. So, I wasn't allowed to do it. I ended up proposing a project based on a problem I didn't care about, but they did approve it.*"

"*I graduated in 2002 during a recession and nobody was hiring. I ended up working in a flower shop, 'designing' corsages and wedding bouquets until April 2003. Then D2M, a boutique product design firm in Mountain View, California, took me in as a part-time volunteer. I volunteered there in the mornings with Kathy Davies, who now teaches at Stanford. She was incredible, I'm so glad she took me in. I eventually transitioned to full-time employment and stayed there for six years. There were really experienced mechanical engineers, industrial designers, and product designers working there. I learned my craft from all of these senior people who knew their stuff. They were graduates from Stanford, MIT, Berkeley, and they knew everything about getting a product to market.*"

"*One thing I appreciate now is Andy Butler, the CEO, was very transparent about how the company worked. Every month he sat us all down and he'd show us, 'Here's how much money we made. Here's where that money comes from in terms of clients. Here's what we need in order to do this.' I started to understand the financial aspects of how product design consulting businesses work. I knew how much margin we charged for everything. Years later when we created Catapult Design, it wasn't really that daunting to me. I understood how product design companies worked from both an employee and client perspective, and the skills that you needed, and process. I understood that you needed to diversify income between multiple clients. Andy taught me that just by having those monthly meetings.*"

"*D2M was nice, and everyone was senior, and I learned a lot, but I wasn't motivated by the types of products we designed. We worked for a lot of power tool companies—Porter Cable, AIMS, Craftsman—and we did a lot of toilet stuff with Kohler. I just felt really jaded creating these things that were, in my mind, useless stuff that would sit on store shelves for a few months and then be replaced the next season with a new design.*"

Heather soon found an outlet for her ambitions with the volunteer chapter of Engineers Without Borders-USA in San Francisco. "*At the first meeting I attended, I met other who didn't feel like they had a place in the typical EWB projects. Most EWB projects are related to civil engineering, water engineering, or structural engineering. There was a strong outlet for mechanical engineers, electrical engineers, software engineers, or product designers. Only one other person I met in that first meeting was familiar with the appropriate technology movement. I was probably the only person in the room who had actually read about the history, or knew about Schumacher, Practical Action, and all that stuff. When the chapter president called for ideas, it was the first time in my life that I vocalized an opinion about something. And because I was the only one in the room with opinions and ideas, and because I also worked at a design firm and knew how products were built, I became the head of what we later called the Appropriate Technology Design Team (ATDT).*"

In 2006, when Heather started ATDT as an initiative of the San Francisco chapter of EWB-USA, it seemed like a perfect use of her talents and ambitions—assisting volunteer chapters to design solutions for developing communities. However, she quickly ran into hurdles.

"*The biggest challenge that I had with the volunteer model was the lead times for all projects. They were insanely long relative to a professional environment. In my day job, we could get a Craftsman tool banged out in three months, whereas it took us several years to even come up with a very basic design for a vertical access turbine. The long timelines led to a lot of volunteer turnover, so there was no knowledge being retained in any of this work.*"

For example, "*the Darfur Stove project would have a build day, and they'd be swarmed with people eager to join the project. They spent a lot of time re-educating everyone on what they're doing. No, you can't use titanium, or no, we're not going to have steel rollers. And they'd spend more time fighting with people about why they couldn't access titanium in Darfur than they would actually get stuff done.*"

Heather had run up against the limits of volunteer organizations. "*I had a double life by then. By day I was at D2M, but while I was at D2M I was answering a bunch of EWB emails because we had 400 volunteers and six projects running at the same time. I was running a design firm while also working full time and it was starting to blur together. It became all consuming. At one point I was biking to work thinking 'I love this stuff so, how can I just make the stuff I do after work my full-time job?' And I thought, 'Why isn't there a design firm that does this?'*'"

Heather presented a plan to EWB-USA to create a dedicated product design arm of the organization called Catapult Design. After a positive reception from the EWB Board of Directors, she put together a budget and a pipeline of proposed projects. But instead of a salary, she was tasked with fundraising for the group, and was also discouraged from charging EWB chapters for her services. So, in 2008, she withdrew Catapult from EWB-USA and established the firm as a separate nonprofit organization.

"*It's funny, at the time people would say, 'oh my God, that's so amazing' about Catapult's founding story. But for me it was extremely incremental thinking. It wasn't innovative by any means. I was working at a design firm already. I was running this little volunteer design firm at night. It just made sense that you would eliminate one so you could do the other. I already knew how a design firm worked, so building that knowledge wasn't a stretch of the imagination. Starting a company didn't feel like a risk to me; it was a necessity in order to do what I wanted to do with my career.*"

Catapult's first major project was the gari roaster, a roasting machine for a grain commonly used in Africa. Catapult also contributed design work to the Tippy Tap handwashing stations, the Hippo roller water transport product, and Zola Electric's mobile app. In addition, Catapult worked on consumables, such as Living Goods fortified food porridge. Over time, Catapult evolved toward service design, in particular in the financial technology arena, assisting solar lighting companies with product and service design. Catapult has since collaborated with Simpa Networks in India and Mastercard to increase access to financial services globally.

After eight years, Heather began to realize that consumer products, by themselves, were not solving poverty problems. "*People can't afford products for the most part unless you're making new lids for jerry cans or something with low cost and high utility. We quickly came to understand that most people can't afford hardware. Even if we try and make it as affordable as possible without sacrificing quality, a $10 or $15 item can still be considered a major expense*." Heather also observed that the entrepreneurs promoting products, and the funders supporting them, often acted based on anecdotal evidence at a remove from lived experiences."

"*When I tried the Hippo water roller in Mountain View, I went to go get water from a little pond near our offices. There was a slight incline to the pond and after I filled it up, I could not get it out of the water. It was so heavy—and this was a very subtle incline*."

Heather gradually became dissatisfied with product design aimed at developing countries. "*I feel guilty that we're creating these products without knowing or understanding all aspects of the problem. Catapult worked on a poop excavation device for pit latrines and I've never used a pit latrine off the reservation. I don't know anything about that space. Catapult worked on a prototype in Nairobi and discussed a long-term pilot with a local sanitation company to test it in multiple locations. It's a good test, but the story being told is that these tools are solving the global sanitation crisis when the reality is, this prototype that we built is not going to suddenly solve the global sanitation crisis*."

"*It speaks to the fact that people gravitate towards good stories and Band-Aid solutions because it's too hard to think about the underlying issues. They're extremely complex and our brains don't work well with complexity or abstraction. We like tangible things that we can see and understand*."

"*In Navajo, if I just think back to the families who can't afford electricity or who can't get the IHS or the NTUA to come out to them to deliver their water service, it's because they have no money. They have no influence. Because there's 60% unemployment on the reservation, there's no jobs, there's no local economies, people just sell what they can on the road sides or in the flea markets, and that's what they rely on for income*."

Heather has recently stepped back from the global development stage, and is turning her efforts to supporting entrepreneurs on the Navajo Nation. "*I am starting a co-working and business incubator on the Navajo Nation, because we have such a large population of sole proprietors who don't see themselves as 'entrepreneurs.' They're 'just' the fried bread makers, the burrito maker, the woman who cuts hair out of her trailer, or the guy who goes around to the chapters and picks up soda cans in his truck who drives all the way to Flagstaff to trade the cans for money. All these people are running informal businesses in order to make ends meet. If they could formalize those businesses, and get to a point where they felt comfortable saying they run a business and pay business taxes, then the tribe would benefit from all of those small businesses. They would reduce unemployment and generate tax revenue*."

"*I know that entrepreneurship is a driver of economies. As someone who's created a business and saw how it transformed me and empowered me, there's so much benefit in just that. The challenge now is remembering and recognizing that the entrepreneurial mindset is not embraced by all. On the reservation if you take a risk and you don't succeed, that could ruin you. When I created Catapult, I always knew that if I didn't succeed, I'd just going find a job. People don't have that luxury in some of the most impoverished places*."

5 Chantal Iribagiza

Abstract

Chantal Iribagiza was born in Rusizi in the far southwest corner of Rwanda. At a very young age, Chantal already knew that she wanted to become a development professional, having been inspired by SNV, a Dutch foreign aid agency that built homes in Rwanda. Chantal earned her engineering degree in Rwanda, has worked as an engineering with Living Water International, the World Bank, and the University of Colorado around the world on rural water services. Chantal's career has included innovating new methods to monitor rural water supplies, as described in Chaps. 1 and 3.

E. Thomas, *The Global Engineers*, Sustainable Development Goals Series, https://doi.org/10.1007/978-3-030-50263-8_5

Chantal Iribagiza was born in Rusizi in the far southwest corner of Rwanda, on the border with the Democratic Republic of the Congo. At a very young age, Chantal already knew that she wanted to become a development professional, having been inspired by SNV, a Dutch foreign aid agency that built homes in the late 1990's for survivors of the 1994 Rwandan genocide against the Tutsis.

"*SNV was helping thousands of homeless widows and orphans in Rusizi get back on their feet after the genocide. The houses that SNV was helping build were more than just homes, they were a sign that things would get better and life would go on. SNV was helping restore hope in times of dire need, and to me that was the most important work I had ever seen.*"

Chantal was educated at girls boarding schools where she excelled, receiving an award at the age of 15 for Best Performing Girl in "O" Level, presented by the Imbuto Foundation, an initiative of First Lady Janet Kagame, which promotes girls education among other objectives.

"*My decision to become an engineer started when I was 15 years old. That is when you get to choose a specialty, after finishing the part of secondary school where you learn everything: geography, history, humanities, and science. I was good at Math and Physics and I had just received an award that boosted my confidence around those subjects. So, I decided to focus on them for the rest of secondary school, and from there it was clear to me that the next thing would be to study engineering. It was also my belief at the time, that engineering offers better employment opportunities.*"

Upon graduation from secondary school at the age of 18, Chantal was recruited to teach Math and Physics to secondary school students in her home district of Rusizi. "*I was teaching 15-year-olds and most of them were taller than me and looked older than me. It was really amazing, though, because it was just a better experience for my students. From their feedback they would tell me, 'I want to be good at Physics, I want to be good at Math. I know I can because you are, and you are not 40 years old.' I was closer to them, they could see themselves in me and be inspired by me.*"

Chantal was accepted by the Kigali Institute of Science and Technology, now known as the University of Rwanda College of Science and Technology, where women comprised a majority of the students in environmental engineering. "*There were more women in my program at KIST than at the universities here in the United States. It is true that more work needs to be done to ensure gender equality and equity in Rwanda and around the world, but it is also true that more needs to be done to uproot misconceptions about what it means to be a woman in a low-income country. Without an understanding of what the challenges are, one risks to cause more harm than good, thinking that they are helping.*"

A talking point among the professional development class is the stated commitment toward empowerment of women and girls. The unspoken assumption is that women are marginalized in many developing countries. Rwanda works against this perception. Rwanda's 2003 Constitution established a quota of at least 30% women in all elected positions. Today, 68% of Rwanda's Parliament and half of the cabinet are women.

A common misconception is that the powerful role of women in Rwanda is a consequence of the genocide. Chantal explains, "*I think that women have always had a place for leadership in the Rwandan culture. When the country was ruled by a monarchy, the second most powerful place was held by a woman: 'Umugabekazi.' That position was usually held by the mother of the reigning king, which is why it is often translated to 'Queen Mother,' even though that it is not its direct translation. If the king's mother wasn't alive at the time their son ascended to power, a different woman would hold the position. Unlike the queen who had indirect power, Umugabekazi had direct power.*"

Chantal's first international development experience was as an intern, then as an engineer, with Living Water International (LWI) in Rwanda. LWI, an international non-profit operating out of Houston, focused primarily on clean water provision in underserved communities.

Chantal worked with staff technicians, district officials, the national Ministry of Environment and Natural Resources, and international LWI staff to implement water projects in Rwanda. Some of these projects included water handpumps.

The typical theory around these kinds of handpumps is that while international donors may support pump installation, local villages remain responsible for operations and maintenance (O&M). The reality over the past decades is more complex. Small communities are ill equipped to manage water supplies, and aid organizations are typically not prepared to support repair costs or services. Meanwhile, the typical role of local governments—providing basic services—is undermined by these projects.

This is the most common model for international support of basic service delivery in developing nations, and generally calls for finite funding and timelines of typically a few years to deploy, maintain, and monitor interventions such as handpumps, latrines, or energy systems. Usually, the impact of these interventions is directly evaluated by implementers. In some cases, funding may be available to employ health epidemiologists or development economists to run randomized controlled trials in order to rigorously evaluate whether the projects are improving environmental, health, or other outcomes. However, even when a positive impact is observed, the majority of these environmental service interventions receive the support of implementers for only a few years. As a consequence, there is increasing evidence that many of the services provided in developing countries have failed to consistently deliver services.

In rural sub-Saharan Africa, where hand pumps are a common technology, 10–67% of improved water sources are non-functional at any one time, and many never be repaired. Problems maintaining the functionality of water pumps result from social, logistical, and technical issues such as the breakdown of community management structures, insufficient human resources to provide service, and lack of spare parts. Community-led activities commonly include creating user committees to determine access arrangements, water fees, and maintenance provision. A community caretaker may provide service and repairs for small problems. For significant problems and pump failures, communities contact a pump mechanic, often one affiliated with the local water district.

Under the contract model, the community or water district hires an organization to manage O&M. The organization may make predetermined, periodic service visits, ranging from once a month to a year, to provide service and fix identified problems with water pumps (the circuit rider service model). Alternatively, communities may call for service as needed (the ad hoc, or nominal, service model). Importantly, the latter model competes for budget and time with other activities technicians and managers are providing, including water pump installations. A consequence of this is substantial downtime due to problems with identifying and reporting pump failures and lack of funds to pay for service and repairs.

In Chantal's view, "*This typical model is not sustainable. Someone needs to be laying out a long-term plan for these facilities, making sure there are funds for maintenance and ensuring that there is supply chain and technical competency that allow for them to be serviced when needed.*"

In 2014 and 2015, Chantal headed up implementation of a project to install cellular network connected sensors on about 200 handpumps in Rwanda. The sensors monitored handpump use, and were used to determine remotely when a pump might be broken. Chantal's job was to install these sensors, and then use the data to deploy technicians for repairs. The project had a major impact on handpump functionality. Before the sensors were installed, some 44% of the pumps were broken at any given time, and it took an average of about seven months to effect a repair. Adding sensors reduced the repair interval to just 26 days, with the result that only 9% of pumps were broken at any one time.

Since her time as a water service professional in Rwanda, Chantal has earned a graduate degree in environmental engineering at Portland State University, and is presently a doctoral student in environmental engineering at the University of Colorado Boulder, while periodically working as a consultant to the Water Practice at the World Bank. Chantal plans to continue working as an engineer in international development. "*I believe that with the right mind-sets, open and genuine collaboration among all parties involved, transparency and accountability, development professionals can make a difference.*"

6 Jean Ntazinda

Abstract

Jean Ntazinda is a Rwandese development professional, working with national and international non-profits, companies, and governments to bring innovative financing and technical solutions to Rwanda. Jean's career has included helping to develop a nationwide health program in Rwanda, described in detail in Chap. 2.

E. Thomas, *The Global Engineers*, Sustainable Development Goals Series,
https://doi.org/10.1007/978-3-030-50263-8_6

Jean Ntazinda first stepped foot in Rwanda at the age of 12. It was 1994, the aftermath of the Rwandan genocide, and his parents were returning to their homeland after fleeing to Burundi in the 1960s. Although Rwandan, he spoke only Kirundi and had to learn his native tongue of Kinyarwanda.

"*When we arrived in Kigali in 1994, I was the only one of six boys left in my family. Four brothers had gone to fight for liberation and one was left in Burundi waiting for his last salary. I remember the very first night I slept outside my home in high school. I found very different people and some of the boys were a little bit aggressive. The reconciliation was very fragile. It was 1996 and wounds were still fresh. The government was working very hard to establish a legal framework and [encourage] their people ... to respect laws and regulations. Every military agent, or soldier, was moving with a gun and a stick. If they saw a villager doing wrong things, they would beat them. They forced people to respect laws at whatever cost.*

"*After five years people got used to the new system, but by then there was a problem of corruption. After another five years, people started getting professional. There were these development projects and people started seeing that they have common interests. They built villages, neighborhoods, then people would get houses and have to live together. It was a process that isn't over yet—you can still see signs of little gaps in the reconciliation.*"

"*But in Rwanda today, all Rwandans have equal rights. There is no identifying which family you are coming from; everyone can get a job, can go to school. This is the big purpose of the Rwanda Patriotic Front (RPF) doctrine. Making people feel they share rights and they have to be competent to get something—not because you are favored by your family member or from your ethnic group. I understand that maybe there are few people that may think it's not the case, but generally you would say that this is the best era that Rwandans have ever lived.*"

Jean studied Geography and majored in environmental planning at the University of Rwanda. He went to work first as an urban planning officer for the Kigali City Council, and then for the Rwanda Environmental Management Authority (REMA). His assignments included working with international project developers to apply the United Nations Clean Development Mechanism (UN CDM) to programs in Rwanda. The CDM system allowed for the generation and sale of carbon credits to fund environmental efforts. In 2009, the US-based social enterprise, Manna Energy Limited, approached REMA to develop the first-ever carbon credit-financed drinking water treatment program. As Jean explained, "*Manna wanted a local staff that may help in understanding the local context.*"

In 2012, Jean quit REMA and became Manna Energy Limited's Carbon Credit Development Officer. Manna was later sold to DelAgua, a UK-based water test kit company, and Jean became DelAgua's country director. In this role, Jean was responsible for negotiating with myriad government ministries to design a large-scale water filter and cookstove intervention, the Tubeho Neza (Live Well).

"*It was very exciting when we were designing Tubeho Neza, when we were negotiating with the government. We had to test the infrastructure that we were going to provide, bringing on board many government agencies, including REMA itself, that would have to review and validate the program under the Clean Development Mechanism of the UNFCCC and the Rwanda Bureau of Standards, to see that the cookstoves really have enough efficiency and the water testing kits can really purify water at an acceptable standard.*"

An early challenge involved negotiating between the Ministry of Health and the Ministry of Infrastructure for oversight of the cookstove intervention. "*The Ministry of Infrastructure and the Ministry of Health had shared responsibilities: the Ministry of Infrastructure was to promote energy efficiency, but the Ministry of Health oversees the household infrastructure to promote hygienic conditions and sanitation at household level.*"

This turf battle also required navigating the roles of international funders and experts. "*When a government is relying on external funding, ministries sometimes want to compete to have partners and push their agenda very fast. The internationals, they push hard. They want things to happen very fast. And government, they have to follow procedures. There is a minimum level of bureaucracy, especially here, people somehow fear taking quick decisions.*"

Meanwhile, another UK-based organization was busy promoting locally produced cookstoves with the Ministry of Infrastructure. The impasse was resolved after an official from WHO contacted the Minister of Health to reinforce the potential health benefits of the cookstoves.

"*This was a sign that development work should be organized in a way [to ensure] that all stakeholders have full ownership. I realized that the Minister of Infrastructure didn't own our program—and yet it was a national program. The Ministry of Health was fully owning the program because they were highly interested in the health outcome.*"

"*We did great work in a very short time. It was like a military operation. We had a very young team and they were very excited. Most of them were fresh graduates. They were so eager to accomplish something. We reached over a hundred thousand families in three months, dispersing cookstoves and water filters, followed by household education visit to teach them how to use the equipment, to teach them the benefits of changing behavior on drinking water.*"

However, shortly after reaching the 100,000 household milestone, and with a vision of providing a further 500,000 homes with products to improve health, DelAgua ran into trouble. Their investors got cold feet and suspended any further funding. Hundreds of thousands of products have already been procured and on their way to Rwanda.

"*DelAgua had all of the units in the warehouse in Kigali, others on the port of Dar es Salaam. We had to terminate our staff contracts, and [provide] compensation for almost 900 community health workers. We had spent a lot of time and resources on training the community health workers on how to educate the household members on using cookstoves and the water filters. And we had given them a timeline or timetable of visiting households, and were supposed to pay them, so the equipment we gave is useless in the household. Behavior change is not a two-day, two-week work. You need to continue. So, people went back to their habits of not respecting the instructions they were given. When do they fill the upper box of the water filter? How do they cook? Do they cook from inside or [do] they have to put the cookstove outside to cook? Some of them would cook small meals, light meals using our cookstoves and cook the rest of the food inside. So ... we didn't achieve the behavior change the way we wanted. And some of the families stopped using the infrastructure we provided. So that was a very bad side of the program.*"

"*I was feeling so disappointed ... I had invested my life, all of me in this program. Convincing government that the funding is there, [that] we are going to make it successful. And the government was going to lose trust. ... I was feeling like maybe it's my fault. I didn't predict this before. But later on, I understood that investors can decide at anytime to change their mind. So, it became about convincing the investor that the program is working and the government is behind it. We had even managed to attract the interest of the World Bank, who agreed to buy carbon credits from our program.*"

"*That was very hard because every staff [member] would ask me what is happening, what is next? Most of them had new contracts that were still fresh and they were so proud of doing great work.*"

"*This kind of diplomacy is not easy. I have tried to see the positive side of working with international companies, and show the government the efforts we are making. But ... I had to show the difficulties we have and manage to lower the expectations from government, so that when things go wrong, they won't be too much disappointed or too much surprised.*"

Jean is now working across multiple sectors and programs, including assisting Bridges to Prosperity to forge a national partnership. "*Rwanda is hilly. So, people are afraid of crossing rivers, especially during the rainy season when they are flooding. Kids may walk* 10 km *to go to school when they could cross a river to go to a school in less than* 2 km. *Same with pregnant women [and] farmers that want to take the harvest to a market. So those bridges help in improving income in the communities, but also helping people to have access to government services, health education and maybe other services they can get at local administration offices.*"

"*The program is designed in a way that it will be getting a payment per performance. In this way, it is like the carbon credit program of Tubeho Neza. There is an independent research effort to measure impact that is accompanying this program. And we've brought many government stakeholders on board. The Ministry of Local Government will be coordinating districts' participation. They'll be helping in organizing, budgeting planning for districts to accommodate this local infrastructure, and [the] budget for the district contributions. And there is [the] Rwanda Transport Development Agency. They'll be supporting payment for the import duties.*"

Throughout Jean's career, he has straddled these two sides, working with Government of Rwanda ministries on behalf or with international organizations.

"*I would categorize the role of international partners in Rwanda in two areas. First of all, there are those ones that directly support the national budget, and second, are those ones that come to support implementation of a sector strategic plan, but outside of the national budget. The common role is filling in the gaps of interventions that a government cannot cover itself. I welcome them, I think we need them. But they should be limited to some of the areas. If I were in a seat to decide what should be their area of intervention, I would ask them to only help in building the capacity of Rwandans, and to give lasting solutions because most of them, they spend enough resources on something that will not last over five years, and we keep needing them. It's like they arrange to be here forever. Most of them they come, they support designing strategies and so forth, and they will tell you that after five years, the strategy is no longer helpful. They have to come back and design another one. They come to do something that they know will expire in less than 10 years. So, for me, I would design or ask them to come to build the capacity of Rwandans, so that at a certain time we'll no longer have to rely on international partners.*"

"*Bridges to Prosperity (B2P) is going an extra mile in terms of building capacity, because the program is designed in such a way that it is kind of fully owned by Rwandans. If I may say, I would give an example, they managed to [include] a bridge design in the curriculum of our technical college, so the engineers we are using in building the bridges have studied that at college. Rwandan engineers build all of the B2P bridges, they know everything. They are getting hands-on skills, how to build that, how to make an architecture design, and they are coordinating the program. So later on, if we have maybe another need of a bridge, we'll have people to design that, and we'll have people to build that.*"

"*I hope that people in Rwanda and internationally will ... start thinking that to be successful you don't have to be a politician. You can be a businessperson or an engineer. And I hope that our international partners can be proud of what we've done, but then leave. And we become business partners, we trade.*"

7 Avery Bang

Abstract

Avery Bang is an American civil engineer, dedicated to leveraging engineering and social business toward reducing rural isolation. She has travelled to over 80 countries as the President and CEO of Bridges to Prosperity (B2P), a Denver-based non-profit that designs and builds pedestrian footbridges. Avery has mobilized capital toward reducing rural isolation and poverty, using innovative financing and engineering approaches, described in Chap. 1.

E. Thomas, *The Global Engineers*, Sustainable Development Goals Series,
https://doi.org/10.1007/978-3-030-50263-8_7

Avery Bang is a bona fide superstar in the development community. Featured in the IMAX film *Dream Big: Engineering Our World*, speaker at various TED events, and recipient of an honorary doctorate at the age of 29, she has travelled to over 80 countries as the President and CEO of Bridges to Prosperity (B2P), a Denver-based non-profit that designs and builds pedestrian footbridges.

Isolation caused by lack of transportation infrastructure affects almost every facet of life for the rural poor. Without adequate transportation access, families cannot access schools, health-care, employment, or local markets to sell and buy goods. The World Bank estimates that nearly a billion people worldwide lack access to an all-season road within 2 km, illustrating the scope of the problem, and the challenge of addressing it at scale. Under Avery's leadership, B2P has constructed more than 320 footbridges in 22 countries, an infrastructure intervention that is cost-effective, durable, and relatively simple to scale. An economic impact evaluation of B2P's footbridges in Nicaragua found a nearly 36% increase in labor market income attributable to the access provided by the bridges.

Avery was born and raised in Iowa, and in her youth was not exposed to either great wealth or poverty. Many young Americans first encounter international cultures and low-income communities during college spring break or semester abroad wanderlust. While studying Civil Engineering at the University of Iowa, Avery travelled to Fiji and immersed herself in the culture.

"*The dynamic between the Indo Fijians and the Fijian population was immediately evident. I started to kind of dig into why the Indo Fijians had something of a second-class citizenship and began to question how and why. It's really evident. You drive down most roads outside the Capital area in Suva and on one side of the road were Fijian households and on the other side were Indo Fijians ... both sides demonstrated a different level of wealth.*"

In Fiji, Avery volunteered with the Breast Cancer Foundation. *"Most of my volunteer time was spent sitting at little booths handing out pamphlets. We also made field visits where we started conversations about the importance of early detection, but given the somewhat taboo subject, specifically in the Fijian culture, I was learning far more than I was able to teach, even when we did reach the communities. But getting there was in fact just as big of a challenge. We would start out in the morning in a bus, van, or if we were lucky, our own car, but as we neared the communities, we were almost always on foot. It was during this last mile of our journey that access became such an issue; the rivers being too high [provoked] the impossible decision—to either turn back or risk our safety to reach those women."* Access to these communities was intertwined with seasonal rains and the volatility of the rivers. *"It didn't occur to me that connectivity could be a root cause of poverty, but within the lens of healthcare access, that lack of connection was very obvious.*"

"*One of those days I was in an area where the Fijian government and the New Zealand government had partnered to put in a pedestrian bridge.*" A result of this intervention which Avery describes was that the connected communities became a tourist destination. "*You kind of imagine these structures are not that complicated. I had a hair-brained idea, I was like, 'Well that's an interesting project. Maybe I could do that for my honors project and maybe I can figure out how to build a bridge for an isolated community. That would be really interesting.'*"

In 2006, Avery got in touch with the founder of B2P, Ken Frantz, a successful retired developer. At the time, B2P was a small organization with an annual budget of about $120,000. Within 18 months, Avery was installing a bridge in Peru in concert with four other University of Iowa undergraduate engineering students. As a result of the experience, Avery went to graduate school at the University of Colorado Boulder, and studied pedestrian bridge design with Engineers Without Borders-USA Founder, Professor Bernard Amadei.

While there she discovered "*a major gap of understanding globally in the space of last-mile infrastructure.*" The existing literature and practices emphasized structural design but disregarded the importance of local soil and rock conditions. "*How do you understand the condition of the soil? How do you understand the complexity of the rock? Those questions were far more challenging than how to design what went between the two sides of the river.*"

During the first years of Avery's leadership of B2P, she identified an opportunity to shift from a private donation-based model to one that engages construction firms and professionals. "*My idea was to get the private sector to pay for this work—I saw an obvious win-win. If you imagine an employer who wants to recruit and retain my peer group of millennials—I had a strong feeling they were going to be able to offer a value proposition that was more personal than just writing a check for the soup kitchen or the local United Way, because that's not a differentiator.*"

B2P's cable-suspended footbridges are well suited for many communities worldwide. Deforestation, erosion, unmanaged rivers, flood events increasing in severity and frequency as the climate changes, and a general lack of heavy machinery all motivate a straightforward solution—largely standardized pedestrian suspension bridges. This approach drives the productive engagement of volunteer engineers from major engineering corporations to contribute to bridge constructions during short-term trips.

Today, B2P's annual budget of over $6 million per year includes several million dollars of support from over 60 corporations, including Parsons, Balfour Beatty, WSP, Bechtel, and other major engineering and construction firms. These firms "*give their employees a sense of the big picture when they give back.*"

"*[When I first started the Corporate partnership model], I ... specifically targeted companies who already understood infrastructure, ideally bridge infrastructure, because there was a natural market alignment, so we did not have to explain why a bridge is important. They already understood the importance of physical connection. It was a win-win for us as well, because at that time I didn't have any other engineering staff; so while the financial contributions grew our organizational efforts, their technical expertise also built up our core engineering and construction capacities.*"

Avery's personal motives have evolved during her career. In the early days, she says, "*I could name any number of the kids at a particular community or a host-mom's best meals. The human element was a clear driver—'If I don't do this and I can't get this money raised, or if I can't get this design done or if I can't get figure out how to get cable procured,' then a family would go without connection. That resulted for me in more burnout because I just couldn't let it go. Over the years, that passion transitioned into more of a pragmatic realist perspective.*" Avery now is comfortable embracing her entrepreneurial drive. "*Even if it's not a big company or a known brand, I'm building an organization. I feel like I've been given this chance to prove that I have the entrepreneurial chops to design and construct something that will have great impact.*"

"*I've always been deeply uncomfortable with the power dynamic of white privilege in general, but critically also within my own staff. To think about my team, I was always deeply uncomfortable by people calling me the boss. ... My role happens to be fundraising and designing and building a team and helping to orchestrate all the resources, but I've always believed deeply in the collective power of a group of individuals brought together to achieve something greater than any one of them could accomplish alone.*"

While Avery values her technical training, she has always been more motivated by what bridges can do, rather than what bridges are—in particular, the positive impacts the bridges can bring for communities and governments.

"*The part that's challenging and interesting is the system. How do you get the money and resources where they need to be? I had a tipping point where I realized that the lack of a bridge alone was not the problem. The root cause of these communities being isolated is not because there's not a technical solution or even a system solution; it's because there's a lack of capital being mobilized to actually solve these problems. If you think about the rural context, the constraint that I saw was, like, there's something in the financial marketplace here: these projects are too small for the national fiscal agents, yet too large collectively for the local government to be able to pay for within existing budgets.*"

In 2016, Avery decided that B2P's business model was stagnating, and that new opportunities might be beyond the reach of her skills and network. She sensed that there were bigger opportunities available if B2P could access global financial markets, but she didn't know how to do this. Avery went to her board and said, "*'I think you guys should either hire another leader or help me re-tool. [B2P needs] someone who has a financial background, understands the capital market, and critically understands big aid, bilateral, multinational government funding. I had a sense that I was not the person that's going to get us to where we need to go without that knowledge. If you look at really the magnitude of the intractable problem we're trying to solve, it's a rounding error. I'm not in the business to keep myself employed.*"

B2P's board decided to allow Avery to take a sabbatical to pursue an MBA at the University of Oxford. At Oxford, Avery led the university's early stage venture capital group, and networked with individuals and organizations operating at the highest levels of both the financial and social enterprise sectors. It became very clear to her that "*There's a moment in time ... as a social entrepreneur that you are in and you're popular and you're in the cool crowd and you have to hit right when you have your idea.*"

Success in this context is governed by gatekeepers—the TED Talks, PopTech, Echoing Green, Mulago Foundation, the Aspen Institute, the Davos World Economic Forum, and Skoll—an escalating and self-dealing network that selects winners. As with many pseudo-meritocracies, where you came from and who you know can matter more than your experience or ideas. When you have both, however, a leader's vision can accelerate. "*I really struggle with the idea that any one group of people not based in the place where they're trying to make deep change is the most qualified to be able to select who is doing the best work, but I do think they have a role. Once you're in, the biggest benefit, I think, is that you're in company with other people doing what you're doing at scale.*"

Avery left Oxford with a new set of skills, a network, and a big idea: "*Our role is to mobilize capital, ethically.*" Avery stopped thinking about each bridge as a separating fundraising and project management effort, and started looking at bridges as a *"portfolio of work."* "*You start to imagine what would happen as opposed to thinking about incremental improvements. What if we actually solved the problem in this district? What if, in Rwanda, everyone actually had basic services? That starts to look like a very different proposition.*"

Today, Bridges to Prosperity has committed to overseeing the construction of over 350 footbridges in the next five years, connecting over 1.1 million people in Rwanda, and meeting the entire identified need. The Government of Rwanda has contributed 40% of the total costs, with foundation and corporate donations accounting for some of the balance. However, the real financial innovation is the introduction of outcome-based funding to this portfolio, or as Avery describes it, "*Getting paid based on results, where the funding effectively pays for the proven outcome.*" The outlines of this approach have been demonstrated in other sectors, typically as "development impact bonds." These have had limited success. Avery elaborates: "*An impact bond within its own right is an individualized, highly sophisticated financial structure, which requires a lot of structuring, so there's a cost to that, a lot of individualized monitoring, and evaluation in order to trigger payment. Overall, the risk is all in one organization. It's not diversified, and so if you think about the outcome payers or buyers as well as the upfront capital investors, there's only so much you want to put in terms of eggs in one basket. They are limited in capital size.*"

Instead, Avery is thinking big. To support B2P's performance-based funding, she has linked up with a variety of other implementers, researchers, funders, and governments to aggregate capital, diversify risk, and lower transaction costs. These contracts, in turn, create revenue streams that can be debt financed.

This career path—from volunteer student to multinational CEO—has offered Avery a perspective on the proper role of Western engineers, and professionals in working to address global development challenges: "*There are needs for Western development professionals that have deep technical skills and are capable of being agile and working under pressure. I think an engineering background is of the highest utility of any kind of background, as it relates to training your brain to have it not only identify problems, but to solve them. But imagining how your little widget, and I include bridges as a widget, are going to meaningfully change the lives of people, I think, has got to be checked at the door. You have to start to imagine yourself as part of a much bigger team. You have to be imagining yourself as working in response to, and in service of, building that local technical talent and supporting government services.*"

"*Our privilege gives us access to resources that are just not available to the development professionals that are our colleagues overseas. In the way the current capital market is structured, it's highly unlikely my colleagues will have access to the same amount of philanthropic resource here in the United States that I will. So that's a big part of our role ... as I believe it is incumbent on us to be thinking about how we can mobilize capital to solve these problems.*"

Doris Kaberia

8

Abstract

Doris Kaberia is a Kenyan expert in food security and pastoral livelihoods. She has worked with Kenyan and international partners to improve food security, drought resilience, and economic prosperity, most recently as the leader of a large-scale development program in northern Kenya. Doris' work is also shared in Chap. 3.

Doris Kaberia saw her first running river as an adult. "*My siblings and myself had never seen a running river when we were young, because we don't have any rivers that flow in the region where I was born, in Eastern Kenya. The most running water we saw was run-off from the rain, especially when there's a lot of rain.*" One of 14 children, Doris was the first girl from her primary and high school to go to university.

E. Thomas, *The Global Engineers*, Sustainable Development Goals Series,
https://doi.org/10.1007/978-3-030-50263-8_8

"My mother was very intelligent, and she had a lot of wisdom, even though she never stepped into a formal school. When we came home in the evening we would go and get water in a 20-L jerry can. We used to walk about 3 or 4 km away from home. We would line up for about an hour and fill the jerry can and bring the water home." Doris has experienced the risks female children face while seeking this precious resource: "*I narrowed escaped a rape by a drunkard while returning home with a 20-L jerry can full of water*" she explained.

"*When we brought the water home, my mum ... would ask, 'What were you taught today? I want to see your book.' And somehow, somewhere, she figured out that when it's a tick, it's correct, and when it's an X, it's wrong. And so, she would say I want tick, tick, and no X. So, she really motivated me to study because she used to say, 'If you come with X, then you'll tell me whether you didn't eat last night.' Because she had not gone to school, she had this drive of having her children go through school. My dad had only gone up to seventh grade, and then bandits came and stole the cows. His parents could not afford to take him to school because their cows were stolen.*"

"*My father also was a champion in education. He really wanted his children to go to school, and he advocated for girl child education. I remember when I left primary school and went to secondary school, my dad would come, like, every two weeks just to visit and see how I was doing. And if he had problems with school fees, he would go to the principal, or maybe you call them superintendent in the US, and explain his situation. 'I don't have school fees, but I don't want my daughter to be sent out of school because she's going to miss on classes. I want her to remain and listen to the teachers and go through the school, but I will bring the money.' And he gave a date and he honored the date.*"

"*I finished high school and I went to Egerton University to do a Bachelor of Science in horticulture. I know how to grow food, vegetables, and fruits both tropical and temperate. We also learned ornamental horticulture; we learned issues to do with landscaping, building of greenhouses, but also, most important, post-harvest management—when you grow food, how you market, how you do processing. We did a number of food science courses; we also did some engineering courses, especially to do with layering and surveying, technical drawing—one and two where we had to use T-square to project; for example, if you want to build a greenhouse, the aerial view, the side view, the side elevation.*"

"*I finished in 2001, but we did not graduate until 2002, because the graduation ceremonies had to be presided [over] by the president, the sitting president. So, the president had a schedule, and if he was not available during your year you had to skip graduation. So, I graduated in 2002, and I went home and asked my dad to give me 15,000, which is about US$150. So, I came to Nairobi to do job hunting—in Kenya we call it tarmacking.*"

"*I was looking at the NGOs that I knew about from home. When I was growing up, there was an organization called Plan International. They used to have projects in my home area and indeed in my school. And I used to see a lady called Christine who used to ride a motorbike, and I was really interested in riding a motorcycle as well. So luckily after I got a short-term job with a flower farm. I would work there on Monday and Tuesday, then on Wednesday and Thursday I was busy hunting for jobs in Nairobi.*"

"*I heard about this organization called the Sustainable Agriculture Community Development Program (SACDEP). It's a Kenyan local NGO that was started by someone called Joseph Mutura, who used to work for Plan International. I went to SACDEP, and when I saw the Executive Director, I told him, 'I have just newly graduated from college, I have heard this is what you're doing, and I think I can make a contribution to your organization. I'm just asking for an internship first, then you can test whether I can make a real contribution.'*"

After six months, Doris secured a full-time position. "*The question I was asked by panels was, 'Do you know how to ride a motorbike?' But I didn't even know how to ride a bicycle, let alone a motorbike, because where I come from it's very hilly—people don't buy bicycles there. So, I answered, 'I think I am a quick learner. After this interview if you give me the motorcycle and a trainer, I'm going to go and learn, and tomorrow morning you'll find me riding.' They decided to give me the job.*"

"*I was given a project area in Gatundu where it's very densely populated, and each of these farmers had a very small piece of land, less than an acre. And if you have four children, this one acre is divided into four. They subdivided to a level that is no longer economical for use. My job was to train them on how they can translate this small piece of land into an economically viable piece of land, so that they will be able to afford their own food security within the household—what we call own household consumption—but also save extra for the market and be able to afford to pay things like sugar, rice—anything that is not produced locally, but at the same time pay for their bills like health and also the school fees for their children.*"

"*One of the program elements was water. We looked at harvesting rainwater from their roofs. We connected the roof with gutters, and then we built a ferrocement water tank, underground water tank, metal mold water tanks for collecting water. Then we had to use this water for multiple uses. So, we could use water for small kitchen gardens. And most of the time what we [grew] was what we call commercially viable crops like tomatoes, sukuma wiki, spinach, dhania (Coriander) green peppers, red peppers. And then we linked the households to markets.*"

"*But at the same time because the land is small, we also had to look for innovative ways of keeping livestock. Instead of keeping a cow that requires you to have three acres to grow the fodder, we had to look at alternative milk-producing animals like dairy goats. We also had beekeeping, setting [up] the apiary to the harvesting, and to processing of the honey. We could do fish farming, training people on how to do the lining of the pond, we get fish fingerlings, and then they [identified] whether it is tilapia, and then we showed them how to feed the fish.*"

"*We also had to think about household energy. People had run out of trees on their land, and had stopped using and cooking traditional, very high nutritive-value food like Githeri, which is a Meru word for beans, and maize. They were cooking rice or cabbage because it required a lot of fuel and they did not have that firewood. So, we introduced biogas with help from some experts in Rwanda, because they could use the cows' manure. When we did the first three biogas plants, it went viral, and all the farmers wanted even to pay their own money to do biogas.*"

"*While I was there, a team of professors from the University of Wisconsin came to visit. They were looking for places to go in Kenya to find innovative strategies of working with HIV and AIDS-impacted people who don't have big pieces of land. These people were taking the HIV treatment drugs and then taking supplementary vitamins, but they were not recovering their body immunity because they were compromised by their poor nutrition. So, it was an effort to improve nutrition among these people.*"

Doris was then invited to attend the University of Wisconsin at Stevens Point. "*When I first arrived in Wisconsin and it was snowing, I thought that everybody cared for me because people thought that I would fall, especially during freezing rain, and everybody wanted to grab me in the stores and everywhere, but also that was very scary the freezing rain. And then when it came to driving, I had to do a driving test. And I was very pregnant so looking at the blind spot was a challenge, and I failed the first test.*"

"*The culture shock was a real thing to me. The first time I showed up in college I thought all the American students looked the same. The food was different. People were used to cold food, but in Kenya we are used to hot dishes. My host family was Tom and Judy, and they really took care of us. From day one when I landed there, they took me to the stores to buy warm jackets, to buy the essential items that I needed because I didn't know that I needed a winter coat and I didn't realize that it was freezing cold. I thought that it was just the normal cold, I didn't know that it would go below zero degrees Fahrenheit.*"

"*They also invited us over for Easter. I got to learn about egg hunting, American Thanksgiving turkey, American Independence Day, hot dogs. We were always together as a family. They also trained us and taught us [about] things to avoid, like credit card [sellers] coming to your door.*"

Both of Doris' children were born while she was in Wisconsin. "*I went to Lamaze classes like any other woman. I was friends with some of the faculty who were pregnant at the same time. When the children were born, we would take them out to the parks. I was juggling my master's degree and motherhood. I was at one point interviewed by national television during Mother's Day to explain how it felt to be a new mom in a foreign country.*"

"*One day, I gave a presentation in a class on international resource management. Afterwards, the professor came to me and said, 'I wish we could have more African students, we could enrich this class a lot.' And I said, 'Yeah, I have one African student in my house, do you need one?' So, he said, "Yeah, sure, bring him in the afternoon." So, I called my husband, and he put the baby in the stroller and came for lunch. That's how Elijah also got to complete graduate school.*"

"*We also learned about timekeeping. When you plan for a meeting here in Kenya, you say we are meeting at 9, but people show up at 10. But I learned in college with other American students, if the lecturer doesn't show up by 15 min after the start of class, you can just leave. Here in Kenya, you would wait until the hour is over.*"

"*After I finished, I went to work for a very rich gentleman who supports Kenya and South Africa. He had a foundation, but the foundation had no staff. So, I showed up at his office, introduced myself and said I could work for him. And he said, 'You are a lioness. You think that I can give you the foundation, the foundation has no employees. So, if you need a job, you'd have to come and interview with the board.' The next day, I went and interviewed with the board—10 people—and they gave me the job. I had to write my own job description and give myself a title.*"

"*With the foundation, we would make regular trips in the US and around the world including to Kenya. But sometimes I thought these trips were stage-managed. So, I decided I needed to move back to Kenya to really make a contribution. It wasn't easy, because the kids were Americans and, at the time, they could not be dual citizens with Kenya. So, we would have to get travel visas all the time.*"

Back in Kenya, Doris started working as the Food Security and Livelihoods Coordinator for Save the Children, a UK-based non-profit. She was posted to northern Kenya, where food security was an emerging concern.

"*It was scary because the first Catholic nuns to be kidnapped in Mandera town, a place called Elwak, were kidnapped when I was there, and taken to Somalia by the Al Shabaab. My life was saved by a local district commissioner, who had some intelligence that we should not travel to the field on certain days. I had young kids at home, so after six months I couldn't take it [any] more and told them I don't think I should save other people's children and not my own. That week my child got rotavirus. Rotavirus is pretty dangerous —my child was so sick he had to go to the hospital, and I wasn't there. I quit. But, I count myself a very lucky person because when I resigned, I had to give a one-month notice. Then CARE advertised their livelihood sector manager position. I applied, interviewed for it and got it. I was only jobless over the weekend. The livelihood sector position had a $10 million portfolio on disaster risk reduction, water and sanitation, agricultural value chains, Disaster Risk Reduction and climate change, financial inclusion, Natural Resource Management and Environment.*"

These experiences qualified Doris for her biggest job yet—as Chief of Party for a five-year, $35 million program funded by USAID and the Swiss Development Corporation. The Kenya Resilient Arid Lands Partnership for Integrated

Development (Kenya RAPID) program focuses on improving resilience in the arid counties of northern Kenya. "*Kenya RAPID is a multi-donor, multi-stakeholder, multi sector program. When you look at water, you don't just look at water as a commodity for human consumption, you look at water as an enabler for other sectors as well. You look at water and ask, 'what is the nexus between water and education?'*"

"*Without water in schools, I don't see the learning environment being very comfortable, so it's an enabler for education. And if the kids do not have safe water, it compromises their intellectual capacity. If they have problems with diarrhea or they have malnutrition issues due to consumption of unsafe water or poor sanitation, then they don't get to develop mentally.*"

"*I believe that the development challenge in the water sector globally requires more than one actor, because if you look at the Kenya context, for example, about 41% of the Kenyan population is actually relying on unsafe water sources. And if you look at urban and rural [areas], the disparity is big. We have a higher percentage of urban population having access, and 50% or less in the rural areas accessing safe water. National government estimates [for safe water] are about a hundred billion shillings every year, but all government and donor funding available in Kenya [amounts to] only 40 billion shillings per year. So, we have a 60-billion-shilling gap per year for basic water services.*"

"*So, bringing in the private sector also is very valuable. Only 3–7% of all the funds that are invested globally for the water sector come from development partners. The rest come from the public and private sector. So, we play a catalytic role in bringing in the private sector. I am not an expert on the private sector, but I quickly learned that when the private sector wants to invest in something, they're asking for what is the business case. How am I going to make this a commercially viable product that can sell on the market? Where is the market, where are the market-driven solutions?*"

Doris is working in the arid north of Kenya, where drought is a recurring issue. "*The reason why we are working in arid and semi-arid areas is because they account for 70–80% of Kenyan landmass, and only 10% of the population—about 4 million people. While it is only home to 10% of the population, it accounts for almost 40% of the agricultural GDP because of the livestock. All the meat that we consume in Kenya is coming from those pastoral range lands, but also it accounts for 10% of the overall GDP.*"

"*We are working in the counties in the north where water access is very limited, and climate change is very evident. Every two to three years we have a major drought. When a drought happens, people lose their lives due to drought, some of them [due to] malnutrition. Also, when it rains, there are some places in these counties like Garissa and Turkana, where flooding happens. So balancing water harvesting, recharging of the aquifers, and planning for the water is very, very important.*"

"*When a drought happens in Kenya, water scarcity is priority number one, priority number two, priority number three. When you talk about food security, you can't produce any food without water. And when drought happens, millions of people are in need of food assistance.*"

"*If you go to talk to people who are over 80 years old, and ask them about historical droughts, it used to happen every five to seven years. And they can tell you very clearly which years those were. But now it is happening every two to three years. It is hotter now, and the amount of rain that these counties are receiving is actually less than 400* mm *per year. In terms of temporal and special distribution, there is also variability. The rains are coming much later, and when they come, they're not spatially distributed well the way it used to be before. One place may flood, and another place sees no rain.*"

"*The livelihoods and the kind of livestock they used to have, the goats and the cows, can't survive anymore because they can't trek for longer distances to get water. But the camel can stay up to four days before it gets water, and it can trek for a much longer distance. So, we have seen goats die, we have seen all these pastoralists losing a large herd of their livestock. And when you lose your herd, it is like losing your money in your bank. In America, you keep money in the bank, so it's like somebody coming to rob your bank account. You have nothing to pay your bills, you can't take your kids to school, you can't pay for your health care.*"

Kenya RAPID's efforts spread across a consortium of donor, government, and non-profit and private sector institutions. *"We are working with the private sector, county water ministers, county chief officers, water directors, the engineers that designed water programs to look at what water sources they have, calculate the water demand and project it for the future depending on the percentage of population growth, and then also factor in multiple uses of water. When we are looking at water demand, they have to look at demand for human consumption, water demand for agricultural production or productive uses. We look at water demand for livestock because these are pastoral communities."*

"*When it comes to digging boreholes, we have seen water levels going down. We have also seen poor water quality. So, as much as you want to drill a borehole as an engineer, as much as you try to work with a hydrologist to tell you that there is water here, you also have to look at the water quality.*"

"*Part of our approach [involves] partnering with the National Drought Management Authority (NDMA) to install satellite connected sensors on all of the strategic drought response boreholes in northern Kenya. This allows NDMA to monitor the functionality of those boreholes, so that during drought periods everybody can ensure that the water is running for people [and they will] not be affected due to lack of safe water.*"

"*The water trucking approach that is normally taken during drought when a borehole breaks down does not benefit the local communities. It benefits rich people, and cartels that find ways of ensuring borehole X is not functioning, so that borehole Y can supply water to those communities. Sometimes [this involves] cutting of the pipelines, or vandalizing some of the equipment that supplies the water. As water sector actors, we have to find ways of breaking these cartels. The cartels and Al Shabab are successful because, as these effects of drought happen in these communities, the youth are recruited because they don't have another source of livelihoods. They're promised big dollars that they have never seen in their lives. They become militants and they just attack to get paid.*"

"*We also have interesting elements of accountability and transparency. In the water system we want to create a transparent and accountable governance structure. So, we have worked with other private sector partners to bring in prepaid water meters that allow a very transparent revenue collection. When people pay for water, they don't use cash, they use a magnetic chip connecting to Safaricom MPesa. They are able to buy a token, they're able to buy water credits, and then they are able to go and buy water. That also reduces the queuing time for women to wait at water points.*"

Increasingly, efforts to increase resilience to climate change are at the forefront of development priorities. *"Even though we are being affected by climate change now, we are not the biggest contributors of global emissions. I do think it's important for the big emitters to accept that their actions are contributing to climate change. Whether we talk about fossil fuels, whether we talk about the fact that people are not planting trees, whether we talk about the contribution of air travel, or vehicles, or other causes of ozone gases, they have to accept first and acknowledge that they're contributing to climate change.*"

"*I think we can all make a contribution. If a US citizen takes responsibility, a Canadian citizen takes responsibility, or a Kenyan citizen takes responsibility, all in our own way—as Martin Luther King said, "You sweep your compound, everybody sweeps their compound, and everywhere will be clean.*"

The role of international professionals and development agencies in Kenya is evolving. "*Kenya was a British colony. I would say there are still remnants of dependency syndrome. The Kenyan government or the Kenyan NGOs still feel that we have to be reliant on foreign assistance, when in fact we could use that foreign assistance to catalyze locally available resources, to bring in new knowledge and technical assistance, to be able to be self-reliant. The journey towards self-reliance has to start with us and realize that there are local solutions, but these local solutions can be informed by the external world."*

"You will find a local community not repairing a borehole, even though they might have the money, and instead waiting for a USAID-funded project to come and do the repair. Or the Kenyan government talking about having a 60-billion-shilling funding gap for our water plan, when we could think of better local solutions. The Kenyan Pooled Water Fund is looking at local pension funds to invest [in]—like government bonds in the water sector."

"I think the role of development partners like USAID, or private sector companies, is severalfold. First, there is a funding gap, and bringing in funding resources is a major contribution. Second, Kenya still needs technical assistance including training, capacity building, and introduction to new technologies. For example, the Famine Early Warning Systems Network (FEWS NET) is a major contribution. Third, we need to partner with research institutions to improve best practices. Without these partnerships and opportunities, I wouldn't know about the work in Ghana, or in Rwanda, or in the United States, that can help me here in Kenya."

"As water sector experts, we can work with engineers and develop new technologies that are appropriate to the local context. For example, when we started Kenya RAPID, SweetSense, an American company, had a hand pump sensor to monitor water supplies. But they quickly realized that what was needed was a deep electrical borehole sensor."

"I am quite optimistic that one day every Kenyan will access safe water and sanitation, but without more acceleration it may take a long time. I still have to learn more, especially on how resilience and climate change connect to the water sector. Water is becoming even more scarce, and resource-based conflicts are increasing in Kenya and our region. I feel very, very strongly that we have more to learn, and that new technologies can also make a contribution to addressing the impact and the effects of climate change."

9 Petros Birhane

Abstract

Petros Birhane, is an Ethiopian agricultural engineer and disaster relief expert. He has worked in Ethiopia, South Sudan, Pakistan, and Indonesia helping communities to recover from natural disasters including tsunamis and droughts. He is now the leader of an internationally funded effort to improve basic water and sanitation access in the lowlands of Ethiopia. Petros' work is also shared in Chap. 3.

E. Thomas, *The Global Engineers*, Sustainable Development Goals Series,
https://doi.org/10.1007/978-3-030-50263-8_9

Petros Birhane was born in Geldid in rural eastern Ethiopia, in 1973, the sixth of ten children. "*We had two springs and during my childhood, as with the rest of Ethiopians, we used to collect water. Usually, we collect water in the early morning. It's two, three kilometers one way. Out of the two spring sources, one of the spring sources, we were told, 'Don't go to that one.' To me, I think it was closer, but quality wise, I realized no. Quality wise, it has, I think maybe minerals, iron, soap—it doesn't taste good. We always go to the furthest one. I know about the challenges associated with access to safe water sources. I know what it means.*"

"*For us, depending on the age, we were given a different size of jerry cans, but I always carried five-liter jerry cans. That's how I started to know the importance of water. We always wait for our turn in the queue to collect water from scoop holes, which as not even an improved water source.*"

"*We were 10, and my father was determined to send us to school. The nearest school was five or six kilometers away, and we had to walk every morning, every day to that school. I remember, we were always given pencil and pens, but we didn't have school bags. We carried those pens and books with our hands. When I came back, I always lost at least either the pencil or pen, and got in trouble with my family.*"

"*I always wanted to study engineering, and I was motivated and encouraged by the engineers working in the natural resources department at that time, and the Rural Road Authority. I studied Agricultural Engineering. That was the only engineering course by then in Alemaya University.*"

"*After graduation, I worked for the Oromia Water Mines and Energy Resource Development Bureau, East Hararghie Office. I had a chance to witness the challenges the community is having to access improved water sources. The eastern part of Oromia is known for droughts and critical water shortages. During my childhood, I also had a chance to witness two or three drought famines. In the place where I was raised, Girawa, I recall seeing many rural people displaced by the famine. I know what it means, what a humanitarian situation is, and what humanitarian responses are even from my childhood.*"

"*After working for almost four years with the government office, I said, I have to join an organization that works for humanitarian responses. I joined CARE International. Luckily enough, I joined the Field Office in Grawa—the place where I was raised. This gave me a chance to work with my colleagues for three years to support the community I came from. I witnessed the advantages and the benefits. I had a chance to see the impacts of our work and the happiness of the community I came from. That's a big reward to me, and then I got motivated for a bigger cause. I said, I have to look for better opportunities to expand my contribution, and I joined the International Rescue Committee (IRC) in 2003 in the West Harargie Field Office. The Program I was managing had intensive WASH interventions in response to the 2002/2003 drought in the area.*"

"*Our work with IRC was mainly focusing on improving access to safe water sources through borehole drillings, electrotechnical installations, and system constructions. We reached over 150,000 people to [help them] get access to improved water sources. We drilled more than 30 boreholes in the area.*"

Petros then moved to Addis Ababa, where he joined the International Rescue Committee (IRC), which focuses on refugee settings. "*Most of our donors were UNHCR, BPRM from USAID, European Union. Compared to the humanitarian response in the host community, I had a chance to again witness and understand contexts and needs, and design responses when it comes to refugees. Refugees are displaced, they don't know the social context, they are new to the environment, so they are vulnerable.*"

After years of working on domestic drought response, funded by international donors, Petros became an international development professional. In 2005, he assisted the tsunami response in Banda Aceh, Indonesia. "*[The area] was one of the most hit by the 2004 tsunami. I went there during the post-tsunami recovery phase, [an experience] that expanded my exposure to humanitarian response. Then, in 2006–2007, I was deployed to South Sudan to provide short-term support through the IRC in the area called Northern Bahr el Ghazal.*"

These experiences convinced Petros to continue working in the international humanitarian system. "*In 2009, I decided to become an expatriate humanitarian professional with the IRC in South Sudan. IRC was the technical lead for the water, sanitation, and hygiene component of the USAID-funded Health Program. Then, in late 2010, I moved to Pakistan to support the emergency response to the Indus river flooding in one of the most impacted areas called Sindh.*"

Petros moved to Pakistan for nearly two years, leaving his family behind in Ethiopia. "*I was there when Osama Bin Laden was assassinated. I was in Islamabad. For us, expatriates were a target and security was a big issue. All our movements were accompanied by local security officers.*"

Returning to Ethiopia, Petros joined the Global Emergency Response team with the IRC—committing to deploying internationally within 72 h' notice. Petros was deployed within Ethiopia and again to Pakistan. In 2013, Petros joined UNHCR as a WASH Officer in Liberia, a posting that gave him another chance to support Ivorian refugees. After working for almost four years as an international humanitarian expert in different countries, Petros returned home to Ethiopia, joining his family in late 2013, and assumed a Project Director role with IRC Ethiopia for the USAID-funded WATER Project. Today, Petros is the Chief of Party for a $27 million USAID-funded WASH program with DT Global, working in lowland areas of the country.

"*In Ethiopia, we have been hit by recurrent droughts in the past starting from the one in 1985, which was one of the worst. Compared to the capacity of our government, the need for humanitarian response usually has been higher. The contribution of humanitarian agencies in countries like Ethiopia is critical to fill gaps. Even today, there are a significant number of humanitarian agencies and developments agencies supporting the Government of Ethiopia and communities in humanitarian and development contexts.*"

"*In the last couple of years, Ethiopia has made significant improvements in health, education, food security, and economic growth. Regardless, the country is still experiencing humanitarian and development challenges resulted from numerous factors including recurrent draughts, flooding, poor agricultural practices, land degradation, and high population growth.*"

"*The challenges we have are complex. They include social, environmental, economic, engineering, financial,* etc. *Coverage of basic service is still way, way lower than standard. For the water sector, we have been working on improving access to water supply services in the country for the last couple of decades, but can you imagine, we don't even have a national standard on water supply system design for rural areas.*"

"*Because of this lack of standards, each project is designed using the experience of the implementing organization. They use their own organizational experience and their educational background. Currently for the project I am running, we developed standard design guidelines to be used by our partners to make sure that what we are doing meets the technical requirements and specifications set both during the design stage and the construction stage. We want to take this experience to the national level. We are working to support the Ministry of Water, Irrigation and Energy (MoWIE) to have this standard established.*"

"*The same is true for solar. Solar is peaking, and the government has a huge interest and plan to promote the use of solar water pumping both for domestic and irrigation purposes. There is a huge gap by partners, even private suppliers, [regarding] properly specifying the right solar pumping equipment for specific needs. Due to this, many solar pumps, which are already installed, failed immediately after installation.*"

"*We are working with the MoWIE, with the Ethiopian Water Technology Institute, well-known solar pump manufacturers and also with private suppliers and consultants at the national and regional level to improve proper use of solar water pumping in the country. Together with the Ministry and other key partners, we supported the development of solar design guidelines.*

Currently, we are also working with MoWIE and the Ethiopian Standard Agency to support the Ministry in having a quality standard for use at the national level. These efforts will contribute to having sustainable water supply services in the country."

Long-term sustainable operation of water supply services has emerged as a challenge in Ethiopia, even as national and international entities continue to drill more boreholes. "*15, 20 years back, the operation and the maintenance were handled mainly by the respective government offices. But over time, the attention given to operation and maintenance has been declining, and the emphasis and attention now is shifting to building new systems. This might be related to limited resources and due to a lack to attention.*"

"*In rural parts of Ethiopia, water is key. For example, if you ask the Somalia Regional Water Bureau about the top three priorities of the region, you get answer: 'Priority one, water, priority two, water, priority three, water.' Therefore, sustaining water services requires increased attention. Currently, O&M of rural water supply services are handled and managed mainly by community-based management committees. This is a great initiative, but the reality is showing that this is not enough, and alternative approaches and modalities are required, including use of the private sector.*"

"*The country, at the national level has great policies, strategies, and guidelines, but these are not cascaded well to regional and community levels. This requires huge attention and advocacy.*"

10 Dan Hollander

Abstract

Dan Hollander is an American hydrological engineer and former United States Agency for International Development (USAID) foreign service officer. Dan has worked around the world including in the West Bank, helping build peace between Israel and the Palestinians through engineering projects. Dan now leads an international program studying water and sanitation services that is described in Chaps. 1 and 3.

E. Thomas, *The Global Engineers*, Sustainable Development Goals Series,
https://doi.org/10.1007/978-3-030-50263-8_10

A hydraulic engineer's idealistic vision of a vocation in water—foundational to life and society—can seem a world away from a corporate office park, and the realities of domestic water resources management.

Dan Hollander was one such engineer, working his first day job developing flood management plans for real estate developments and suburban municipalities, and volunteering with the Chicago Professional Chapter of Engineers Without Borders-USA on a water storage project for a community in Guerrero, Mexico.

Dan earned his undergraduate and graduate degrees at the University of Illinois in civil and environmental engineering with an emphasis in hydrology. The EWB-USA Chicago chapter introduced Dan to experienced professional engineers working local and globally. "*The EWB-USA Chicago chapter was super impressive. The people on the Mexico project included senior managers at major global engineering firms, engineering lawyers, and expert engineers in the Chicago area. I was just starting my first job, and these were middle and upper management folks, and we were all working together on this project. It was a good experience for me, not just to learn about international development, but also to have peers and role models within the engineering community.*"

After a few years, Dan's passion for global engineering outweighed his professional interest in domestic engineering. He googled "international engineering jobs" and found the United States Agency for International Development (USAID) Engineering Foreign Service. Dan applied online, flew to Washington DC for an interview, and within a year, in 2009, was sworn into the Foreign Service, as part of the first wave of new engineers hired by USAID to revitalize the Agency's engineering cadre.

"*In the early days back when USAID was established around the Vietnam War, there were hundreds, if not thousands of engineers. Then in the 1990s, USAID scaled way back and started outsourcing and privatizing a lot of the technical roles—construction, engineering design, oversight, management. The Engineering Foreign Service went from thousands to a handful of engineers by the early 2000s. It was during the George W. Bush presidency that USAID started realizing that they didn't have the technical staff to oversee what was happening in the field. They were using reputable, large engineering firms for construction and design, but they didn't have engineers in-house to manage and oversee the work. To fix this, USAID hired about 20 of us in 2008–2010 as the first new batch of engineers in a long time.*"

After a year of training in Washington DC in learning how the government and USAID work, and learning Arabic, Dan was deployed to Israel, supporting USAID-funded engineering programs in the West Bank and Gaza. "*We had about 400 million dollars annually in water and infrastructure projects. I would either be supporting infrastructure activities like roads, water, wastewater, construction or working on solicitations—evaluating proposals related to engineering construction activities.*"

The USAID/West Bank & Gaza water and infrastructure team included 11 engineers—three Americans and eight Israeli or Palestinian Foreign Service National engineers. Their mission was to build functional, viable institutions to support a two-state solution.

The engineering and construction activities USAID funded in the West Bank was part of a diplomatic and peace-building strategy. "*Infrastructure is a big part of daily life. We built roads work that supported the economy. We developed water and wastewater infrastructure which supported health. We built hospitals, schools and healthcare clinics that supported public services. We worked directly with West Bank Water Departments, helping them with billing, collections, and setting up regulations.*"

These projects helped build trust and legitimacy in the Palestinian Authority (PA). This required careful negotiations both with the PA and the Israeli government. "*If you want to build something you have to both go to the PA because it's in the West Bank and Palestinian, but it's also in "Judea and Samaria", so you have to go to Israel and get their permits too. There was a lot of multiple approvals, multiple import processes, a lot of bureaucracy. Things moved slow and required a lot of time and money.*"

In Dan's second year in Israel, he worked directly with the Office of the Special Envoy for Middle East Peace, as a technical advisor of sorts to the diplomatic process. "*I worked with the negotiating team to support the construction and rehabilitation of public institutions in politically contested areas. For example, a school or a health clinic in a community directly adjacent to a settlement.*"

"*One of the interesting things about being an engineer in diplomatic settings, and perhaps also in development institutions generally, is that you end up being one of the few folks with a strong technical footing. So they'll listen to you. If you tell them that something is technically infeasible or that the Israelis have an objection for particularly good technical reason or health and safety reason, they'll listen to you. That's why they hired you—to be the technical voice in the room.*"

On a temporary basis in his last year, Dan became the acting Director for Water and Infrastructure at USAID/West Bank and Gaza. He was 29 years old, responsible for a portfolio of 1.5 billion dollars in infrastructure activities. Much of this work was managed by the Palestinian engineers. Dan viewed his job as primarily supporting these professionals. "*These are some of the top engineers and they should be the ones making the decisions. I'm not the right person to go in and decide technical decisions or to work directly with the PA, they're much better at that. I was responsible for managing the US contractors and we had to make sure that the engineering efforts were feasible and appropriate. For example, we knew we needed to start working in wastewater. Untreated wastewater was contaminating the aquifers in the West Bank, and making people sick. So USAID decided we would invest in wastewater treatment. Our A&E contracted designed tertiary plants that cost $40 million dollars each. The West Bank didn't have any real wastewater treatment coverage, even primary, in most places. There was no real staff in the PA to manage or regulate these plants, and no one that knew how to run or maintain these plants. There was no land available for it, no clear way that spare parts were going to be available, etc. These designs went all the way to 30 or 50% before somebody pulled the plug on it. A lot of global engineering is making sure engineers, and particularly locally knowledgeable ones, are part of the conversation, and helping to make the right decisions.*"

After supporting the Palestinian Mission, Dan's next assignment was in Thailand, working as a regional engineering officer supporting USAID programs globally. His main job was to help with new program development, or assessment of completed or ongoing programs that were having issues with engineering or infrastructure. In East Timor, he observed a failed USAID-funded rural water distribution project. "*USAID installed small piped systems in rural areas that were essentially surface water catchments and tap stands in a few communities for each system. Pretty soon after ribbon cutting everything broke and nobody fixed it. There were no spare parts and there was no way to communicate to the people who were responsible for maintenance, and nobody knew who was accountable. I went out there to set up a contract for somebody to fix it all. It helped USAID honor its commitment to the communities, but this was not a sustainable solution. The idea was to figure out what type of expertise was needed on a medium-term basis to develop some operation and maintenance processes, not just for initial repairs, but to transfer it to the Timorese government entities to maintain these systems.*"

Unlike his experiences in middle-income countries, Dan was starting to see that engineering alone, in low income settings, was not sufficient to guarantee that basic needs were met. "*We were just providing things without any real long-term vision, or with any support to the overall system that's actually going to sustain it. If you go and give somebody a phone with one charger and the cord breaks, they can't use that phone anymore. And that was essentially what happened with these water systems. One part broke once and then they were never fixed and non-functional. They're made out of either PVC or galvanized pipes. You couldn't buy them except for in the capital. These places were a two-day drive. There was no clear person responsible for them. There's no way it was going to be sustained.*"

After four years abroad, Dan returned to the United States and left the Foreign Service. He spent a few years with USAID contractors supporting water projects, and nearly a year with a congressional advocacy firm where he worked with non-profits, donors, members of Congress, and the State Department to build bipartisan support for investments in the global water sector.

Dan is now the Chief of Party of the USAID Sustainable Water, Sanitation and Hygiene Learning Partnership. "*We are a global project that looks at solutions to the sustainability issues around water and sanitation through non-technical solutions. We are exploring the non-physical systems, institutions, regulations, and people needed to work together to sustain and maintain a service.*"

Dan has seen the differences in the role of engineers across country income levels. "*In development programs supporting middle income countries, the biggest problems with engineering projects I've seen have been gaps in traditional infrastructure design, management, operations, and oversight. You don't necessarily have the right people at the design phase. You don't necessarily hire the right people to oversee the work. Designs or materials selected might make sense for Europe, but aren't sustainable locally, given the reality of local resources, availability of spare parts, and regulations. Many of these gaps occur because locally knowledgeable engineers are not meaningfully involved in the process. In order to make sure that solicitations are designed properly so that the work is performed by locally-minded experts, you have to have the right people keeping an eye on it. In low income settings, the gaps are in public health and basic capacity, and the challenges to sustainability are much more stark. When looking at the context in Egypt versus Ethiopia its really no comparison—your starting point is so much different, and the need to think about sustainability is that much greater.*"

Dan is now working with implementers, researchers, engineers, and engineering students to professionalize global engineering. "*I often advise students about the importance of getting a Professional Engineer license, especially if you are going to work in another culture, another country, another context. It's beneficial to have something that shows that you have that technical background, and you respect the importance of having a technical base. You're going to work directly with a Timorese engineer, a Ugandan engineer, a Palestinian engineer. They all have their local professional engineering license or certification. You can build trust and respect as professionals. You can work together as peers.*"

Dan is particularly proud of his work with the stone cutting industry in the Southern West Bank. "*It was a classical transboundary water issue. Hebron was generating a large amount of slurry that comes out of cutting stone. And this slurry runoff was flowing into Israel, ending up in an Israeli wastewater treatment plant that was not designed for stone cutting slurry. The industry was the backbone of the entire region, and that stone was sold in major markets in Israel. And yet the Israeli wastewater facility was certainly not designed for industrial waste, and was being forced to shut down due to the pollutant load. It was an important issue to solve for everyone.*"

At first, there were indications that the Palestinian businesses would defy any new regulations, and that the Israeli Army would unilaterally go in and shut down the stone cutting industry. Instead, "*we were able to develop an agreement between both the wastewater treatment facility, the municipal government, and all the different Palestinian ministerial entities—and there were a lot of them. We agreed to design and deploy on-site pretreatment, which allowed the Israeli WWTP to handle the effluent that ultimately showed up at its doors. The Palestinian police were given the authority to go in and shut people down in the industrial zone who weren't complying, and they did it! And we accomplished this without shutting down the full industrial zone, or getting into political finger-pointing. Everyone had this initial desire to be firm and not play ball—obviously, there's a long history of conflict and mistrust here. And yet everyone signed a reasonable agreement, made concessions, the economy continued to grow, and the rivers were cleaner, both in Palestinian and Israeli communities. It's not every day that you can build peace as an engineer.*"

Index

E. Thomas, *The Global Engineers*, Sustainable Development Goals Series,
https://doi.org/10.1007/978-3-030-50263-8

CPSIA information can be obtained
at www.ICGtesting.com
Printed in the USA
LVHW060023220720
661236LV00004B/41

9 783030 502621